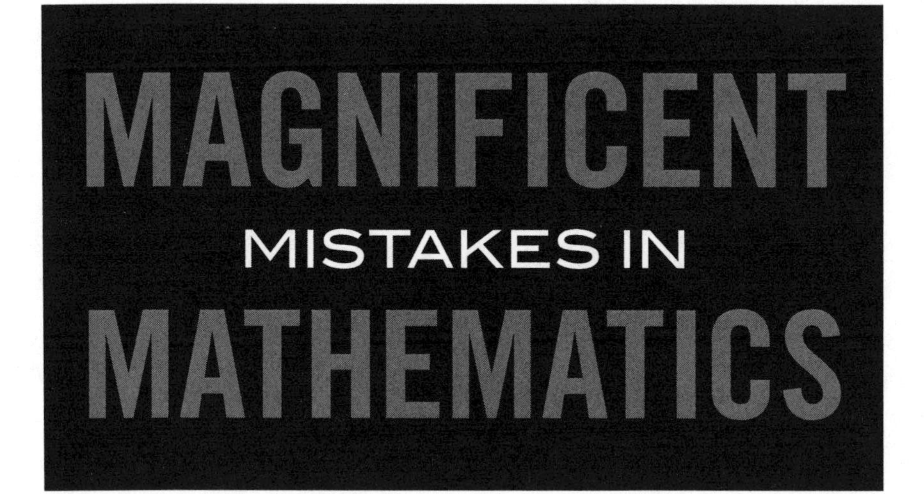
MAGNIFICENT
MISTAKES IN
MATHEMATICS

Alfred S. Posamentier
Ingmar Lehmann

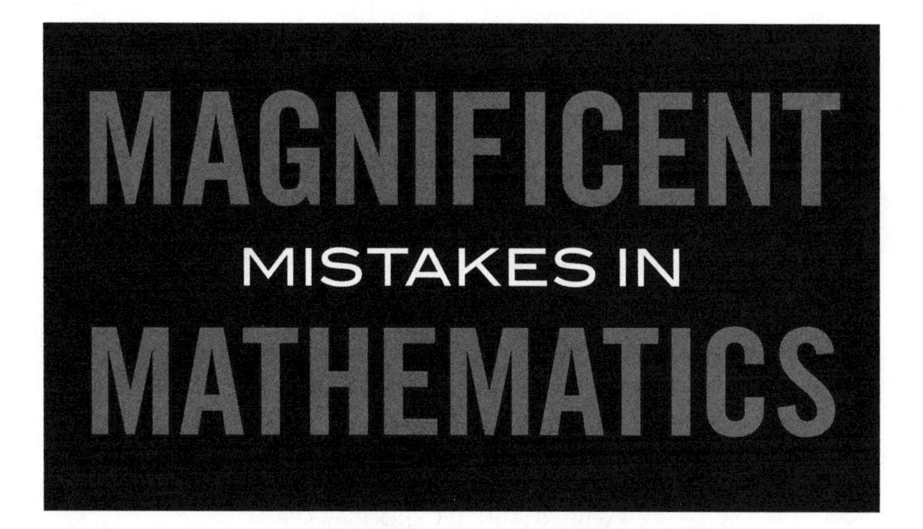

59 John Glenn Drive
Amherst, New York 14228–2119

Published 2013 by Prometheus Books

Cover design by Nicole Sommer-Lecht
Cover image © MediaBakery
Unless otherwise indicated, all interior images (figures and photographs)
are by Alfred S. Posamentier and/or Ingmar Lehmann.

Inquiries should be addressed to
Prometheus Books
59 John Glenn Drive
Amherst, New York 14228–2119
VOICE: 716–691–0133 • FAX: 716–691–0137
WWW.PROMETHEUSBOOKS.COM

17 16 15 14 13   5 4 3 2 1

Library of Congress Cataloging-in-Publication Data

Posamentier, Alfred S.
Magnificent mistakes in mathematics / by Alfred S. Posamentier and Ingmar Lehmann.
pages cm
Includes bibliographical references and index.
ISBN 978-1-61614-747-1 (hardback)
ISBN 978-1-61614-748-8 (ebook)
1. Mathematics—Miscellanea. 2. Errors, Scientific. 3. Discoveries in science. I. Lehmann, Ingmar. II. Title.

QA99.P664 2013
510—dc23

2013012126

Printed in the United States of America

To Barbara, for her support, patience, and inspiration.
To my children and grandchildren, whose future is unbounded:
David, Lauren, Lisa, Danny, Max, Sam, and Jack.
And in memory of my beloved parents, Alice and Ernest,
who never lost faith in me.

Alfred S. Posamentier

To my wife and life partner, Sabine, without whose support and patience
my work on this book would not have been possible.
And to my children and grandchildren: Maren, Claudia, Simon, and Miriam.

Ingmar Lehmann

CONTENTS

ACKNOWLEDGMENTS

The authors wish to extend sincere thanks for proofreading and useful suggestions to Dr. Bernd Thaller, professor of mathematics at Karl Franzens University–Graz, Austria, and Dr. Peter Schöpf, professor emeritus of mathematics at Karl Franzens University–Graz, Austria. Dr. Michael Engber, professor emeritus at the City College of the City University of New York provided some useful suggestions throughout the book. For some meticulous proofreading we sincerely thank Peter Poole. We also thank Catherine Roberts-Abel for very capably managing the production of this book and Jade Zora Scibilia for truly outstanding editing at the various phases of production.

INTRODUCTION

The title of this book can be interpreted in a number of ways. In the study of mathematics, we have all made mistakes along the way. We are not referring to the kind of errors due to carelessness or lack of understanding or even silly errors of notation.

Some mistakes are merely well-hidden subtle errors. Take, for example, the English mathematician William Shanks (1812–1882), who required fifteen years for his calculation of the value of π to set the record for the number of decimal places in 1874. In 1937, in Hall 31 of the Palais de la Decouverte (today a Paris science museum on Franklin D. Roosevelt Avenue), this value of π was produced with large wooden numerals on the ceiling (a cupola) in the form of a spiral. This was a nice dedication to this famous number, but, surprisingly, there was an error. Shanks's approximation was discovered to have contained a mistake, which occurred at the 528th decimal place. This was first detected in 1946 with the aid of a mechanical desk calculator—using only seventy hours of running time! This error in the "π room" on the museum's ceiling was corrected soon thereafter in 1949. The race for accuracy for the value of π is today in the trillions of decimal places.[1]

While we speak of mistakes on public display, consider the clock atop the tower on St. Marien Church, the oldest structure in the town of Bergen on the German island of Rügen in the Baltic Sea. Damaged during a storm in 1983, the clock was restored in 1985 by craftsmen who found themselves with a conundrum. As they were inserting the minute markers on the face of the clock, they found that there was a large and unexpected space between the 10 and the 11. So, they simply inserted another marker to fill the space. Consequently, this may be the only clock in the world with sixty-one intervals on its face rather than the proper sixty intervals. (See figure I.1.)

Figure I.1. Courtesy of Norbert Rösler, sexton of the church St. Marien, Bergen / Rügen (Germany).

There is also a clever story about a standardized test that presented two pyramids: one comprised of four equilateral triangular faces (a regular tetrahedron) and the other a pyramid with a square base and lateral sides that were equilateral triangles congruent to those of the regular tetrahedron. Students were asked to place the two pyramids together by overlapping their mutually congruent equilateral faces, and then to determine how many faces the resulting figure had. Given that the sum of the faces of the two pyramids is $4 + 5 = 9$, and that overlapping two of these faces eliminates them, the "correct" answer was given to be seven faces. It was not until some years later that a student persevered to show that in fact, this was a wrong answer, since by this combination of pyramids, two rhombuses are formed by adjacent equilateral triangles, thus resulting in only five faces. We will take a closer look at this situation in chapter 4. Such errors give us pause to examine even the "obvious."

Our objective in this book is to entertain the reader with a collection of wrong conclusions—or fallacies—that help us to better understand important aspects of or concepts in mathematics. It is through these "mistakes" that we should get a much better insight of and appreciation for of the subject matter. Some "mistakes" can lead to very interesting mathematical ideas. For this reason, we have deemed them "magnificent mistakes." But rest assured that no special mathematical skills are needed to explore with us these fascinating mistakes. We expect the reader to know high-school mathematics and nothing necessarily beyond that.

Sometimes we make simple mistakes that ultimately lead to an absurd result, and we then tend to dismiss the mistake. We know that if equals are multiplied by equals, then the results are equal. For example, if we know that $x = y$, then we also can conclude that $3x = 3y$. Yet, when we have 2 pounds = 32 ounces, and $\frac{1}{2}$ pound = 8 ounces, then does $2 \cdot \frac{1}{2}$ pound $= 32 \cdot 8$ ounces? Or does 1 pound = 256 ounces? Of course not. Where did we go wrong? Similarly, we know that $\frac{1}{4}$ dollar = 25 cents. Then does $\sqrt{\frac{1}{4}\text{ dollar}} = \sqrt{25\text{ cents}}$, or $\frac{1}{2}$ dollar = 5 cents? Again, absurd! Where did we go wrong? When we multiplied the numbers or took their square root, we didn't do that to the units, which would have led us to a correct solution—albeit an awkward one! To make this explanation a bit simpler, suppose we begin with 2 feet = 24 inches, and $\frac{1}{2}$ foot = 6 inches; then, by multiplying the units and the measures, we get 1 square foot = 144 square inches, which is correct!

We can also consider how the "proof" that $1 = 0$ leads us to a most important mathematical concept: that division by zero is not permissible. Follow along as we show this interesting little "proof." We begin with our given information that $x = 0$. We then multiply both sides of this equation by $x - 1$ to get $x(x - 1) = 0$. Now dividing both sides by x leaves us with $x - 1 = 0$, which in turn tells us that $x = 1$. However, we began with $x = 0$. Therefore, 1 must equal 0. Absurd! Our procedure was correct. So why did we end up with an absurd result? Yes, we divided by zero when we divided both sides of the equation by x. Division by zero is not permitted in mathematics, as it will lead us to silly conclusions. This is just one of many such entertaining mistakes that give us a more genuine understanding of the "rules" of mathematics.

These examples may seem entertaining, and they are. Yet through these entertaining illustrations of mistakes a lot is to be learned about mathematical rules and concepts. For example, when we "prove" that every triangle is isosceles, we are violating a concept not even known to Euclid—that of betweenness. When we show that the sum of the lengths of two legs of a right triangle is equal to the length of the hypotenuse—clearly violating the time-honored Pythagorean theorem—we will be showing a misuse of the concept of infinity. Yet it is the unique value of these mistakes—providing a better understanding of the basic concepts of mathematics—that makes these mistakes magnificent. Lest we forget, youngsters—and, we dare say, adults as well—learn quite a bit from mistakes. We expect that through the playful style in which we present these mistakes, the reader will be delightfully informed! We shall also compare mathematical mistakes with those in everyday life and notice what can be learned from these.

We expect that the readers will enjoy these examples, and during this delightful excursion they should appreciate the many aspects or nuances of mathematics that sometimes go unnoticed until they lead one astray. We invite you now to begin your journey through these many magnificent mistakes in mathematics.

CHAPTER 1

NOTEWORTHY MISTAKES BY FAMOUS MATHEMATICIANS

There are numerous conjectures by famous and less-famous mathematicians that have been published over the years. Some of these conjectures have been subsequently supported by proof, while some have been dismissed as mistaken, and others are still seeking verification or dismissal. Yet in all cases the attempts to grapple with these conjectures have moved our understanding of mathematics to the next level. Our journey through some of these conjectures will be to see what it took to verify them and what might have been found to dismiss them.

With a broad overview we can see that some of the greatest thinkers—mathematicians and scientists—have made mistaken conjectures, many of which have led to new discoveries and a broadening of their respective fields. For example, Aristotle (384–322 BCE), one of the most influential and well-known thinkers of all time, has made some mistaken statements. Although one might consider him one of the founders of science, and what he wrote has influenced generations, his errors have opened new fields of thought and study. Here a few of his mistaken beliefs:

- The world consists of five elements: fire, water, air, earth, and ether. The first four are nature on earth, and the fifth fills the heavens.
- Heavier weights fall to earth faster than lighter ones—yet because of Galileo Galilei (1564–1642) we know better.
- Flies have four legs.
- Women have fewer teeth than men.

Aristotle also believed that the Earth was the center of the universe, and that the other observable bodies—such as the moon, the sun, and the planets—revolve around it. His words were quite influential in his time and lasted quite some time beyond. He influenced some of the other great thinkers such as Hipparchus of Nicaea (ca. 190–120 BCE) and Claudius Ptolemy (ca. 100–before 180). Not until Nicolaus Copernicus (1473–1543), Tycho Brahe (1546–1601), and Johannes Kepler (1571–1630) proved that the universe was solar centric did these earlier beliefs lose favor. Originally Kepler believed that the planets traveled on the surface of a sphere—with a circular path—so as to be boxed in by the platonic solids. This was nicely supportive of Pythagoras's beliefs even though they were false! Later on, he corrected his erroneous conjecture by stating that the planets traveled on an elliptical path and described through his three famous laws the nature of the planet's travel along the elliptical path—perhaps one of the greatest achievements in astronomy and mathematics, and one that has solidified his fame forever.

Despite his enlightened thinking, Kepler still had a weakness for astrology. Similarly, Newton had a fascination with alchemy and a deep belief in religion and mystiques of numerology. This latter interest motivated Newton to produce thousands of pages of numerological calculations, which led him to predict that the world will come to an end in the year 2060. Was this a mistake?

While speaking of the earth, consider the most important mistake in the history of America. The Italian explorer Christopher Columbus (1451–1506), through his calculations, was convinced that the western route to India was considerably shorter than that in the easterly direction. This belief was based on the mistaken measurement by the astronomer Claudius Ptolemy, who calculated the circumference of the Earth to be 28,000 km. With this estimation for the earth's circumference, Columbus's conjecture would have been correct. However, the experts in the Spanish court assessed the earth's circumference at 39,000 km, which was arrived at by the Italian mathematician and cartographer Paolo Toscanelli (1397–1482). This calculation was incredibly accurate for the times, as it is quite close to the actual circumference of 40,075 km. Thus, Columbus's calculation was destroyed in the Spanish court. It was also common knowledge at the time that the earth was spherical, and thus Columbus was not trying to prove that it wasn't flat.

Some mistakes are—by today's knowledge—downright silly. The British physicist William Thompson (1824–1907)—perhaps better known as Lord Kelvin, after whom the Kelvin temperature scale is named—believed that there would never be an airplane that would be heavier than air!

The famous Austrian psychologist Sigmund Freud (1856–1939) was always fascinated with the mystique of numbers. Despite his brilliance, he was enchanted with the conjecture offered by the German biologist Wilhelm Fliess (1858–1928) that any number expressible by the combination of multiples 23 and 28 (or put another way that could be expressed as $23x + 28y$), had some special significance in a person's life cycle. He then claimed that many people die at age 51, and found that $23 \cdot 1 + 28 \cdot 1 = 51$. Yet, $23 \cdot 3 + 28 \cdot (-2) = 13$—not usually considered a desirable number! The fact that almost all numbers can be expressed by $23x + 28y$ did not occur to Freud. Another rather silly mistake!

Let us now consider some mathematical mistakes made by some magnificent mathematicians.

PYTHAGORAS'S MISTAKE

Pythagoras of Samos (ca. 570 BCE–510 BCE) is most famous for the relationship bearing his name concerning the sides of a right triangle. Unfortunately, we do not have any of his writings, but there is much attributed to him, nevertheless. Today, we have strong evidence that even his famous theorem was known to the Babylonians and the Egyptians in the special case of a triangle with side lengths 3, 4, and 5 hundreds of years earlier.

A society centered on Pythagoras evolved that was fascinated with numbers, and it was felt that everything could be explained with numbers—namely, the natural numbers: 1, 2, 3, 4, 5, ... The belief was that principles of mathematics were the principles that explained the world. Harmony and nature were to have been explained through these numbers. For example, in music, intervals can be determined by number relationships. In all these instances, he was able to explain the relationships through the natural numbers and their ratios. Wanting to extend this to geometric forms, Pythagoras erred.

One of his society members, Hippasus of Metapontum (ca. fifth century BCE) dispelled Pythagoras's belief that the dimensions of a regular pentagram could also be measured with natural numbers, by discovering that there must be other numbers (later developed as the irrational numbers) to explain the relationships of the various segment lengths of a pentagram (see figures 1.1 and 1.2). This was shocking to Pythagoras, perhaps because this geometric figure was the symbol of the Pythagoreans.

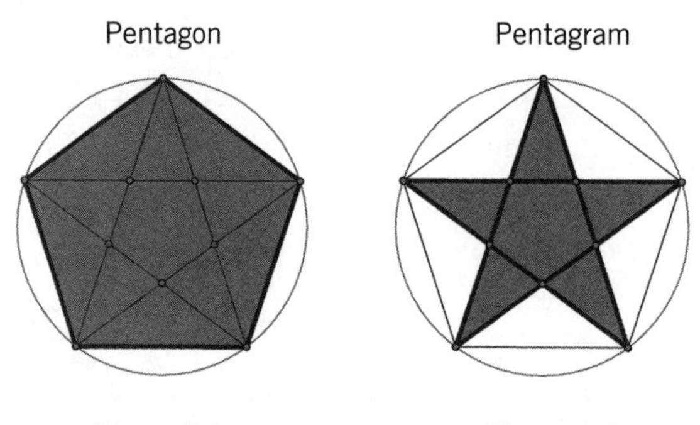

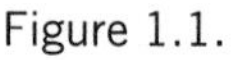

Figure 1.1. Figure 1.2.

The relationship between the segments of a pentagram has some very interesting properties: $\frac{d}{a}=\frac{a}{e}=\frac{e}{f}$ (see figure 1.3).

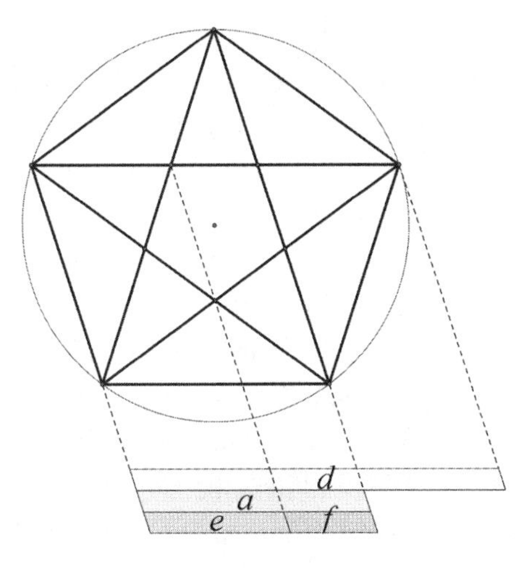

Figure 1.3.

This relationship relates to the famous golden ratio,[1]

$$\phi = \frac{\sqrt{5}+1}{2} \approx 1{,}6180339887498948482045868343656381177720.$$

As a postscript to correcting this mistake of Pythagoras, Hippasus is supposed to have drowned as punishment for this "sacrilege"—extending beyond the natural numbers, which were supposed to define everything!

MISTAKES IN THE ORIGINAL LOGARITHMS TABLES

In 1614, the Scottish mathematician John Napier (1550–1617) published a book on logarithms, which was recognized as particularly significant by the English mathematician Henry Briggs (1561–1630), who, in 1624, consequently published a table to logarithms to fourteen decimal places for the base 10 numbers, titled *Arithmetica logarithmica*. Curiously, Briggs's table of logarithms contained the numbers from 1 to 100,000, but omitted the numbers from 20,000 to 90,000. This void was filled by the Dutch mathematician Adrian Vlacq (1600–1667), who in addition to being a good mathematician was also a clever book merchant. Furthermore, he was particularly important in the popularization of the use of logarithms. His complete table of logarithms first appeared in 1628, as a second edition of Briggs's *Arithmetica logarithmica*, and provided a complete listing of the logarithms of numbers from 1 to 100,000, but only up to ten decimal places.

In 1794, an improved printing of *Vlacq's Table* was published in Leipzig, Germany, by Baron Jurij Bartolomej Vega (Latin: *Georgius Bartholomaei Vecha*; German: *Georg Freiherr von Vega*; 1754–1802), which became the model for all future number tables. In the preface for this publication, the author stated that the first person to find any mistakes in his tables, which could lead to a faulty calculation, would be paid in ducats.[2] Needless to say, it would have been quite astonishing if this table were fault-free. In the course of time, about three hundred mistakes were discovered—and not merely in the final decimal place!

FERMAT'S LAST THEOREM

One of the most famous conjectures is the longtime famous Fermat's last theorem, which states that $x^n + y^n = z^n$ has no (non-zero) integer solutions for $n > 2$. Of course, we know that for $n = 1$ this is trivial, and for $n = 2$ we have the Pythagorean theorem. This conjecture by the famous French mathematician Pierre de Fermat (1607/08–1665) was made in 1637 and was finally proved by Andrew Wiles (and Richard Taylor) in 1995. So for 358 years, Fermat's comments in the margin of one of his books, Diophantus's *Arithmetica*, stood unproven, where he wrote that

> it is impossible to separate a cube into two cubes, or a fourth power into two fourth powers, or in general, any power higher than the second power into two like powers. I have discovered a truly marvelous proof of this, which this margin is too narrow to contain.

Today, some assume that Fermat had a proof for this special case where $n = 4$ and then figured it could be generalized for all values of n. Some number theorists today doubt that Fermat ever did have a proof for all values of $n > 2$.

Along the long path from 1637 to 1995 there has been a series of many mistaken attempts where people thought they found the proof that Fermat considered and others presented proofs of their own invention that also were discovered to have a mistake. The attraction of this 358-year-old puzzle lies in the simplicity of its statement, so that laymen and famous mathematicians such as Leonhard Euler (1707–1783), Ernst Eduard Kummer (1810–1893), Carl Friedrich Gauss (1777–1855), and Augustin Louis Cauchy (1789–1857) attempted to solve this problem—each time with mistakes.

In 1770, Euler showed that the equation $x^3 + y^3 = z^3$ has no solution among the natural numbers. Yet this proof was not complete and required some help in 1830 by the French mathematician Adrien-Marie Legendre (1752–1833). Gauss also provided a correct solution for the case where $n = 3$. We should note that in 1738 Euler successfully dealt with the case where $n = 4$. However, later on it was discovered that the case where $n = 4$ was already discovered in 1676 by Bernard Frenicle de Bessy (ca.1605–1675). After Euler's death there were many futile attempts to solve Fermat's conjecture, notably one by the French mathematician Sophie Germain (1776–1803), who, as a woman living at the time, was forced to publish under the pseudonym Monsieur Le Blanc and set the stage for the proof that for $n = 5$, Fermat's conjecture held true.

Yet, despite the mistaken attempts, these did help further develop algebraic number theory. In 1828, Peter Lejeune Dirichlet (1805–1859) and Adrien-Marie Legendre were able to show that the equation $x^5 + y^5 = z^5$ does not have an integer solution. In this process, Dirichlet made a mistake that was eventually corrected by Legendre. The mistake in Dirichlet's proof meant that it was not complete; but with Legendre's help, it was correctly completed.

This process of proof attempts to support Fermat's conjecture continued, such that in 1839 Gabriel Lamé (1795–1870) proved that $x^7 + y^7 = z^7$ does not have an integer solution, a result that was also shown independently by Victor A. Lebesgue (1791–1875). In 1841, Lamé mistakenly thought that he had proved the general case—that which Fermat stated, since the case for $n = 14$ was put to rest in 1832 by Dirichlet. Finally, in 1847 some top mathematicians such as Cauchy and Lamé were (mistakenly) convinced that the general case of $x^n + y^n = z^n$ was finally proved and were ready to present it to the French Academy of Science. However, Ernst Eduard Kummer destroyed their hopes for fame because he found a mistake in their work.

The drama continued, where in 1905 a Goldmark prize worth 100,000 marks was offered by Paul Friedrich Wolfskehl (1856–1906) for a correct solution of this long-standing problem. This certainly enticed a multitude of mathematicians. On March 9 and 10, 1988, an article in the *Washington Post* and the *New York Times* reported that the thirty-year-old Japanese mathematician Yoichi Miyaoka finally proved Fermat's conjecture. However, in short order this highly vaunted attempt also was discovered to have mistakes.

In 1993, during a workshop at the Isaac Newton Institute at Cambridge University, the British mathematician Andrew Wiles (1953–) over several days presented a proof that appeared to have been finally a correct proof of Fermat's last theorem. However, once again, soon afterward, Nicholas Katz (1943–) found a mistake in Wiles's proof. Wiles, along with his doctoral student Richard Taylor, then spent the next year feverishly trying to eradicate this error. On September 19, 1994, the mistake was corrected and, finally, Andrew Wiles had conquered this famous 358-year-old mathematical challenge. In June 1997, over a thousand scientists gathered at the University of Göttingen (Germany) as the Wolfskehl Prize (worth about $25,000) was awarded to Andrew Wiles, and with this he ended his ten-year journey to the solution of the famous Fermat conjecture. Despite the many mistakes made on the path to success, many interesting by-products in mathematics were discovered, which shows that sometimes mistakes in mathematics can be considered magnificent because they provide some valuable new mathematical insights.

GALILEO GALILEI'S BIG MISTAKE!

The famous mathematician, physicist, and astronomer Galileo Galilei (1564–1642) concerned himself with uniformly accelerated motion for over forty years. His experimental innovation consisted of using the oblique plane, which he was able to use to study the laws of motion and was able to test them from a quantitative standpoint. In 1638, while searching for the fastest connection that a weight would travel between two points under the influence of gravity and without friction, he encountered a mistake.[3] He noticed the time from *A* to *B* along a certain polygonal path was faster than that along line *AB* (see figure 1.4).

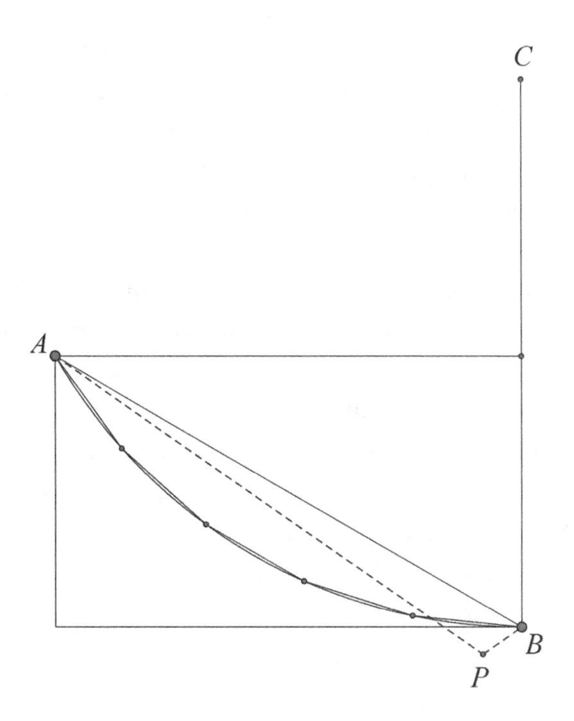

Figure 1.4.

He calculated this speed to make the so-called Galileo pendulum, with which he was able to establish the characteristics of a swinging ball. He noticed that the time of the ball's travel was inversely proportional to the number of vertices of the polygonal path. The more vertices the path has, the less travel time is required. Recognizing that by constantly increasing the number of vertices, the polygon surface will approach the arc of a circle, he therefore conjectured that the arc of a circle must be the fastest curve for the ball to travel rather than a straight line.

What he didn't consider was that the polygon sides do not necessarily have the same length, and, therefore, do not approach the curve of the circle.

Supposing we consider the point *P* as shown in figure 1.4 to be below point *B*, which is something that did not even occur to Galileo, then possibly the speed through the steepness of *AP* will compensate for this segment *PB*. In 1696, Johann Bernoulli (1667–1748) also investigated this problem. He, along with Mersenne, in 1644; Huygens, in 1657; and Lord Brouncker, in 1662, found the solution in *Brachistochrone*.[4] When we look back today, we see the calculus of variations embedded in this situation. Bernoulli also considered the polygonal path as the ideal path, for he felt segments should be longer near the point *A* than the point *B*. This alteration prevented him from drawing the same conclusion as Galileo, since his path would not approach the curve of a circle. So we can see that while the solution that Galileo proposed was a good approximation, it did not give us the exact curve that we would require. Galileo's "false Brachistochrone" is, therefore, an arc of a circle![5]

In figure 1.4, *BC* is the radius of the circle of which the arc is the presumed ideal path. On the Brachistochrone-path, the ball will go from *A* to the goal point *B* much faster than along the oblique plane. In figure 1.5, the dashed line shows this circular arc path.

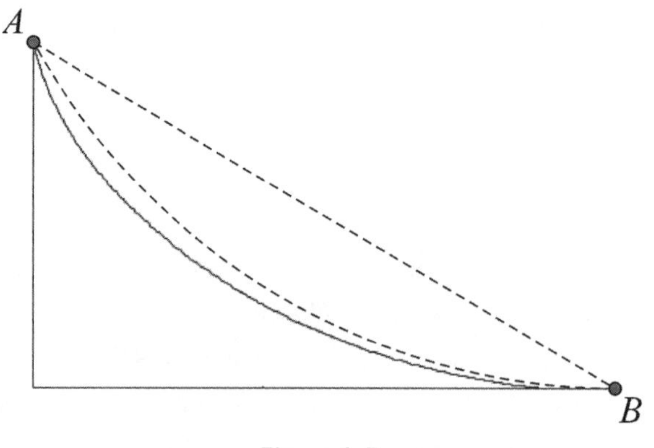

Figure 1.5.

As long as the starting point of the path is higher than the endpoint, and not directly beneath it, then the Brachistochrone—or fastest path—is a curve in the shape of a cycloid. We should note that the low point of the cycloid path can lie lower than the endpoint.

THE HISTORIC MISTAKE MADE BY CHEVALIER DE MÉRÉ

The French nobleman Antoine Gombaud (1607–1684), perhaps better known as Chevalier de Méré, amused himself with games of chance in French salons. He was fascinated with the chances of tossing a six on a single die thrown four times verses tossing a double six on twenty-four tosses of a pair of dice.

However, Chevalier de Méré was mistakenly of the opinion that the two events had equal probabilities, since they are in proper proportion, namely, $4 : 6 = 24 : 36$. With this thinking, he lost lots of money and, desperate as he was, he contacted one of the greatest mathematicians of his time, Blaise Pascal (1623–1662). Pascal, fascinated with the dilemma, included this problem in an exchange of letters with another very famous mathematician, the aforementioned Pierre de Fermat. This correspondence was the germination of a new field in mathematics: probability. Méré's problem, known in the literature as the de Méré paradox had only a minor role in this exchange of letters.

Let's see why Chevalier de Méré's thinking was in error.

The One-Die Situation:

If we roll a die four times, then the total number of all possible outcomes is $6 \cdot 6 \cdot 6 \cdot 6 = 1{,}296$.

Out of these 1,296 tosses, there are $5 \cdot 5 \cdot 5 \cdot 5 = 625$ outcomes that include no six.

If we bet on getting at least one six when tossing a die four times, there are:

- 625 possibilities of losing and
- $1{,}296 - 625 = 671$ possibilities of winning.

Therefore, the probability of not tossing a six on a single die tossed four times is:

$$p = \left(1 - \frac{1}{6}\right)^4 = \left(\frac{5}{6}\right)^4 = \frac{625}{1296} = 0.4822530\ldots < \frac{1}{2}.$$

This shows that our chances of winning with this game are greater than our chances of losing.

Two-Dice Situation:

Tossing two dice once leads to one of thirty-six possible outcomes, namely, all possible outcomes of rolling die number 1 combined with all possible outcomes of tossing die number 2. Thus, if we roll two dice twenty-four times, then the total number of possible outcomes is $36 \cdot 36 \cdot \ldots \cdot 36$ (36 multiplied with itself twenty-four times), which is approximately 22,452, 257,707,350,000,000,000,000,000,000,000,000.

Out of these, there are $35 \cdot 35 \cdot \ldots \cdot 35$ (35 multiplied with itself twenty-four times), which is approximately

11,419,131,242,070,000,000,000,000,000,000,000,000

outcomes without a double six.

Therefore, to get at least one double six when tossing two dice twenty-four times, there are approximately

- 11,419,131,242,070,000,000,000,000,000,000,000,000 possibilities of losing, and
- 22,452,257,707,350,000,000,000,000,000,000,000,000
  – 11,419,131,242,070,000,000,000,000,000,000,000,000
  = 11,033,126,465,280,000,000,000,000,000,000,000,000 possibilities of winning.

Put another way, the probability of not tossing a double six with a pair of dice tossed twenty-four times is:

$$p = \left(1 - \frac{1}{36}\right)^{24} = \left(\frac{35}{36}\right)^{24} = \frac{1500625}{1679616} = 0.5085961\ldots > \frac{1}{2}.$$

This means that the chances of winning with this game are lower than the chances of losing—as the Chevalier De Méré learned the hard way.

GOTTFRIED WILHELM LEIBNIZ'S MISTAKE

One of the greatest minds in philosophy, physics, and mathematics (having been one the codevelopers of modern-day calculus), Gottfried Wilhelm Leibniz (1646–1716) seemed to have made a mistake with the sum of the series: $\frac{1}{1}+\frac{1}{2}+\frac{1}{4}+\frac{1}{8}+\frac{1}{16}+\frac{1}{32}+\frac{1}{64}+\frac{1}{128}+\ldots$, in the way he determined its sum to be 2.

Leibniz did the following to get this series sum:

$$s=\frac{1}{1}+\frac{1}{2}+\frac{1}{4}+\frac{1}{8}+\frac{1}{16}+\frac{1}{32}+\frac{1}{64}+\frac{1}{128}+\ldots \qquad (1)$$

Multiplying both sides of the equation by 2:

$$2s=\frac{2}{1}+\frac{2}{2}+\frac{2}{4}+\frac{2}{8}+\frac{2}{16}+\frac{2}{32}+\frac{2}{64}+\frac{2}{128}+\ldots=2+1+\frac{1}{2}+\frac{1}{4}+\frac{1}{8}+\frac{1}{16}+\frac{1}{32}+\frac{1}{64}+\ldots \qquad (2)$$

Now subtracting the equation (1) from (2):

$$2s-s=s=2+1-1+\frac{1}{2}-\frac{1}{2}+\frac{1}{4}-\frac{1}{4}+\frac{1}{8}-\frac{1}{8}+\frac{1}{16}-\frac{1}{16}+\frac{1}{32}-\frac{1}{32}+\frac{1}{64}-\frac{1}{64}\pm\ldots$$

$$=2+(1-1)+\left(\frac{1}{2}-\frac{1}{2}\right)+\left(\frac{1}{4}-\frac{1}{4}\right)+\left(\frac{1}{8}-\frac{1}{8}\right)+\left(\frac{1}{16}-\frac{1}{16}\right)+\left(\frac{1}{32}-\frac{1}{32}\right)+\left(\frac{1}{64}-\frac{1}{64}\right)\pm\ldots$$

Therefore, $s = 2$.

Why is Leibniz's procedure in question? What assumptions has Leibniz made?

Archimedes (ca. 287–212 BCE) had already considered this series of reciprocals of the powers of 2 and concluded that the sum cannot be greater or less than 2. That would indicate that its sum must be 2. Archimedes thus used indirect arguments to reach his conclusion. He was comfortable with this result, which also trapped Leibniz.

A geometric series $(a_n) = (a_0, a_1, a_2, a_3, \ldots)$ is a series where the ratio q

of two consecutive numbers is a constant. In this case, $q = \frac{a_{k+1}}{a_k} = \frac{1}{2}$. By definition, such a geometric series is the sum of its partial sums.

The series $(s_n) = (a_0, a_0 + a_1, a_0 + a_1 + a_2 + a_3, a_0 + a_1 + a_2 + a_3 + \ldots + a_n, \ldots)$ has partial sums as shown in the following table:

n	s_n				
0	a_0	=	1	=	1
1	$a_0 + a_1$	=	$1+\frac{1}{2}$	=	1.5
2	$a_0 + a_1 + a_2$	=	$1+\frac{1}{2}+\frac{1}{4}$	=	1.75
3	$a_0 + a_1 + a_2 + a_3$	=	$1+\frac{1}{2}+\frac{1}{4}+\frac{1}{8}$	=	1.875
4	$a_0 + a_1 + a_2 + a_3 + a_4$	=	$1+\frac{1}{2}+\frac{1}{4}+\frac{1}{8}+\frac{1}{16}$	=	1.9375
5	$a_0 + a_1 + a_2 + a_3 + a_4 + a_5$	=	$1+\frac{1}{2}+\frac{1}{4}+\frac{1}{8}+\frac{1}{16}+\frac{1}{32}$	=	1.96875
...					
10	$a_0 + a_1 + a_2 + \ldots + a_{10}$	=	$1+\frac{1}{2}+\frac{1}{4}+\frac{1}{8}+\ldots+\frac{1}{1024}$	=	1.9990234375

Therefore, the final sum for s is as follows: $s = \frac{a_0}{1-q} = \frac{1}{1-\frac{1}{2}} = 2$.

Is Leibniz, therefore, vindicated? Not really.

Suppose we now take the reciprocals of the reciprocals of the powers of 2.

$$1 + 2 + 4 + 8 + 16 + 32 + 64 + 128 + \ldots$$

The ratio of two consecutive members of the series is

$q = \dfrac{a_{k+1}}{a_k} = 2$, so that the sum s is given with $s = \dfrac{a_0}{1-q} = \dfrac{1}{1-2} = -1$.

This is quite obviously false. It is quite clear that the sum should be $s = 1 + 2 + 4 + 8 + 16 + 32 + 64 + 128 + \ldots = \infty$, that is, that the series grows continuously or is considered divergent.

The formula for the sum of a geometric series only holds true when $|q| < 1$ (or $a_0 = 0$)—and here this assumption has been damaged.

In defense of Leibniz, it must be said that much of our knowledge of series and sequences was not yet known. This brings us to admire even more so the marvelous work of the earlier mathematicians, such as Archimedes or Euler, who plowed ahead with intuition that still amazes us today.

MISTAKES WITH MALFATTI'S PROBLEM

The topic of packing is one that has fascinated mathematicians for centuries. In 1802, the Italian mathematician Gianfrancesco Malfatti (1731–1807) found a solution to the following problem, which he published in the following year. The problem was to find how three tangent circles can be placed in a given triangle so that the sum of their areas is maximized.

Malfatti claimed that the three circles had to be tangent to each other, and that each circle had to be tangent to two sides of the triangle (see figure 1.6).

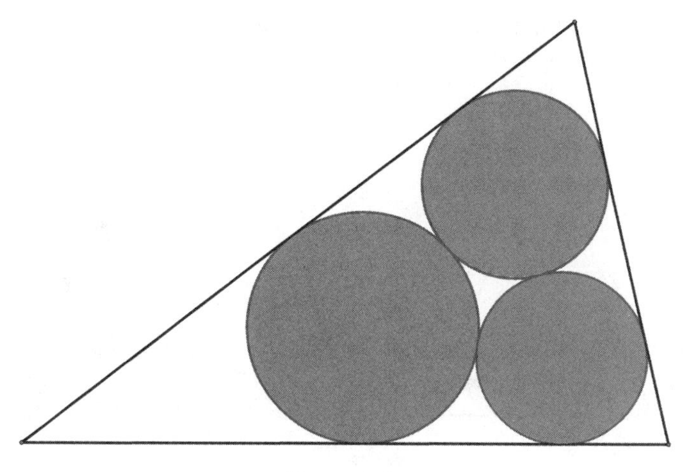

Figure 1.6.

In the case where the given triangle is isosceles, the problem had been tackled by the Swiss mathematician Jacob Bernoulli (1654–1705). For the general triangle, the Swiss mathematician Jacob Steiner (1796–1863) published an elegant construction in 1826.[6]

Numerous other mathematicians played with this problem, such as Julius

Plücker (1801–1868), Arthur Cayley (1821–1895), and Alfred Clebsch (1833–1872), each of whom subscribed to Malfatti's belief that each of the three circles must be tangent to two sides of the triangle. However, in 1929, H. Lob and Herbert William Richmond (1863–1948)[7] discovered something striking. They found that Malfatti had made a mistake. They found that his conjecture didn't hold when it was applied to an equilateral triangle.

When Malfatti's method is applied to an equilateral triangle, about 73 percent of the area of the triangle is covered by the three circles—or more precisely, $\left(\sqrt{3}-\frac{3}{2}\right)\cdot\pi\approx 0.729$. This can be seen in figure 1.7. However, Lob and Richmond were able to find a way to increase the area covered by the circles by 1 percent. To do this, they showed that one of the circles must be the largest circle that can be inscribed in the equilateral triangle, namely, the inscribed circle. The remaining two circles are then placed in the two corners, tangent to the inscribed circle and two sides of the triangle. This configuration covers an area of about 74 percent, or more precisely: $\frac{11\sqrt{3}}{81}\cdot\pi\approx 0.739$ (see figure 1.8). This would then indicate that Malfatti's construction for the maximum area of the circles was a mistake.

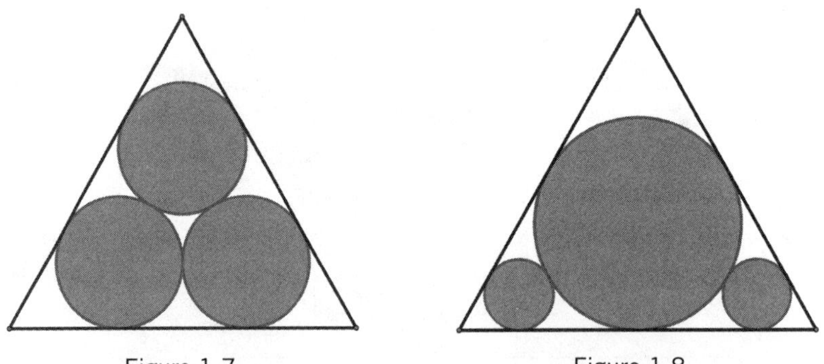

Figure 1.7.

Figure 1.8.

One would think that we have now settled this issue. Not so fast! In 1965, Howard W. Eves (1911–2004) showed that if a long, thin triangle is used, the circles placed as in figure 1.9 encompass a larger area than the configuration of figure 1.10.

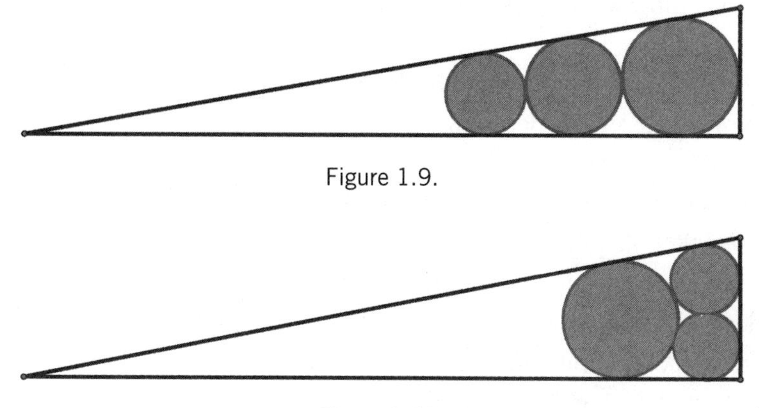

Figure 1.9.

Figure 1.10.

This once again damages Malfatti's claim that the circles had to be tangent to each other.

Then in 1967 Michael Goldberg showed that Malfatti's claim is wrong despite the shape of the triangle in question.[8] The British mathematician Richard Guy (1916–) said that Malfatti simply misstated his problem.[9] The correct solution is that one of the three circles must be the inscribed circle of the triangle, and the other two circles must be tangent to the inscribed circle and to two of the sides of the triangle (see figures 1.11 and 1.12).

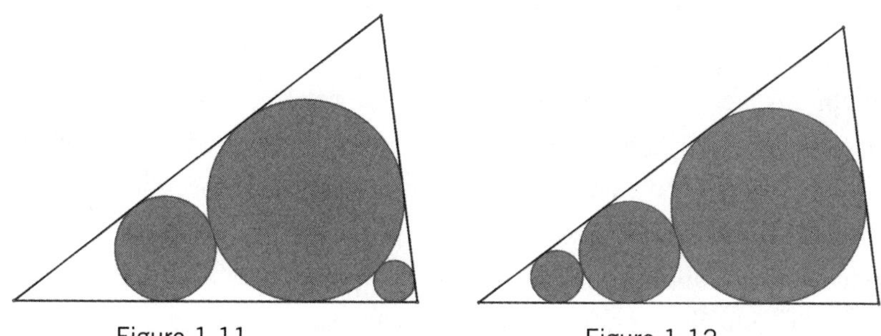

Figure 1.11. Figure 1.12.

A complete proof of this construction did not appear before 1992, when it was published by V. A. Zalgaller and G. A. Los.[10]

WILLIAM SHANKS'S EMBARRASSING MISTAKE

We have come a long way in our determining the value of π, which is the ratio of the circumference to the diameter of a circle. The first few decimal places of π are: π = 3.1415926535897932384 . . .

However, as of October 18, 2011, we have this value correct to 10,000,000,000,050 decimal places, thanks to the tireless work of the Japanese systems engineer Shigeru Kondo and the American graduate student Alexander Yee, whose computer worked on this for 371 days in 2011. Without a computer calculating, determining the value of π was a herculean task. Archimedes was able with much struggle to get the value correct to two decimal places. The German-Dutch mathematician Ludolph van Ceulen (1540–1610) spent years to generate this value correctly to thirty-five decimal places. In 1853, William Rutherford, an English mathematician extended the then-known value of π to 440 decimal places. In 1874, one of Rutherford's students, William Shanks (1812–1882), extended the value of π to 707 decimal places—requiring fifteen years to accomplish this feat.

However, something just wasn't right. The British mathematician Augustus de Morgan (1806–1871) noticed something curious in this value of π: after the 500th decimal place, the digit 7 seemed to appear less frequently. This was quite puzzling, since it was assumed that the digits in the decimal approximation of π would appear evenly distributed. De Morgan couldn't explain this lack of 7s. But his instinct was correct. Shanks made a mistake. There was an error in the 528th place, which was first detected in 1946 with the aid of an electronic computer—using seventy hours of running time!

D. F. Ferguson used the formula $\pi = 12 \arctan\frac{1}{4} + 4 \arctan\frac{1}{20} + 4 \arctan\frac{1}{1985}$ to produce a new approximation to 620 decimal places and discovered the mistake in Shanks's approximation. As far as we can see, de Morgan's conjecture did turn out to be correct in that the digits appear to be evenly distributed in the expansion of the approximation of π. Although this conjecture is not proven, generations of mathematicians have been fascinated with the question of the frequency of the digits in the expansion of π. It continues as a question seeking a definitive answer.

Shanks's mistake has had a rather embarrassing result. In 1937, in Hall 31 of the Palais de la Decouverte—today a Paris science museum (on Franklin D. Roosevelt Avenue)—the value of π was produced with large wooden numerals on the ceiling (a cupola) in the form of a spiral. This was a nice dedication to this famous number, but they used the approximation generated in 1874 by William Shanks, with the error in the 528th decimal place. It was not until 1949 that this mistake on the on the museum's ceiling was corrected.

FOUR-COLOR MAP PROBLEM

The four-color map problem dates back to 1852 when Francis Guthrie (1831–1899), while trying to color the map of counties of England, noticed that four colors sufficed. He asked his brother Frederick if it was true that *any* map can be colored using only four colors in such a way that adjacent regions (i.e., those sharing a common boundary segment, not just a point) receive different colors. Frederick Guthrie then communicated the conjecture to British mathematician Augustus de Morgan.

As early as 1879, the British barrister Alfred B. Kempe (1849–1922) produced an attempted proof, but in 1890 it was shown to be wrong by Percy J. Heawood (1861–1955). Heawood occupied himself for sixty years with this problem, seeking maps to color and knowing that with five colors it was clearly done, but still not convinced that four colors could suffice. Many other subsequent attempts also proved fallacious.

In April 1975, William McGregor in Wappinger Falls, New York, produced a map in *Scientific American* that he claimed required five colors. Yet, in October 1975, Dieter Herrmann showed him to be mistaken by demonstrating that this map could be colored with only four colors.

In his April 1975 column in *Scientific American*, Martin Gardner (1914–2010) claimed that this 110-region map required five colors and was thereby a counterexample to the four-color conjecture (see figure 1.13).

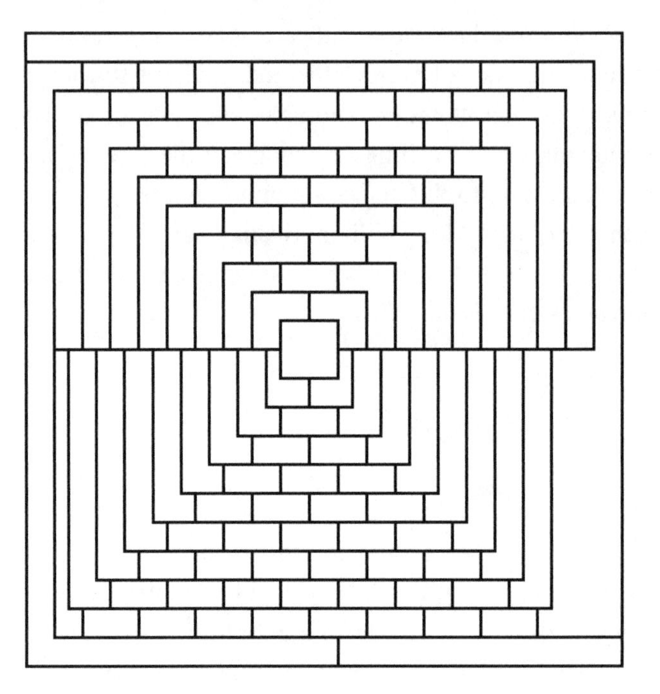

Figure 1.13.

Then, in a subsequent issue, Martin Gardner confessed that his entire April 1975 column was an April Fool's joke on his readers.[11] In fact, Stan Wagon, a mathematics professor at Macalester College, produced a coloring that showed only four colors are needed to color Gardner's map using the computer algebra system (CAS) Mathematica. This is a rare case where a mistake was intentionally produced.

Not until 1976 was this so-called four-color map problem solved by two mathematicians, Kenneth Appel (1932–) and Wolfgang Haken (1928–), who, using a computer, considered all possible maps and established that it was never necessary to use more than four colors to color a map so that no two territories, sharing a common border, would be represented by the same color.[12] They used an IBM 360 that required about 1,200 hours to test the 1,936 cases, and which later turned out to require inspecting another 1,476 cases. This "computer proof" was not widely accepted by pure mathematicians. Yet, in 2004, the mathematicians Benjamin Werner and

Georges Gonthier produced a formal mathematical solution to the four-color map problem that validated Appel's and Haken's assertion.

The attractive aspect of the four-color map problem lies in the fact that it is very easily understood, but the solution has proved to be very elusive—fraught with mistaken attempts at a solution—and very complex.

THE CATALAN CONJECTURE

This conjecture dates back to Levi ben Gershon (1288–1344) and states that the powers 2^3 and 3^2, whose values are 8 and 9, respectively, are the only such powers that result in consecutive numbers. In other words, for $x, y, m, n > 1$, the equation $x^m - y^n = 1$ is true only when $x = 3$, $y = 2$, $m = 2$, and $n = 3$.

In 1738, Leonhard Euler showed that these were the only numbers that would have this relationship among the squares and cubic powers. This was then generalized in 1844 by the Belgian mathematician Eugène Charles Catalan (1814–1894) as for all powers the one cited above is the only one that makes this relationship true. There were many mistaken attempts at proving this conjecture. However, in 2002, the Romanian mathematician Preda Mihăilescu (1955–), after working on this problem for three years, finally proved it.[13]

One might be interested in seeing that this theorem can be extended to other separations between powers as shown in the following chart:

Separation	$x^m - y^n$
1	$3^2 - 2^3$
2	$3^3 - 5^2$
3	$2^7 - 5^3$
4	$5^3 - 11^2$
5	$3^2 - 2^2 = 2^5 - 3^3$
6	?
7	$2^{15} - 181^2$
8	$4^2 - 2^3$
9	$6^2 - 3^3$

10	$13^3 - 3^7$
11	$3^3 - 2^4$
12	$47^2 - 13^3$
13	$4^4 - 3^5 = 16^2 - 3^5$
14	?
15	$4^3 - 7^2$
16	$2^5 - 2^4$
17	$3^4 - 4^3 = 7^2 - 2^5$
18	$3^3 - 3^2$
19	$10^2 - 3^4$
20	$6^3 - 14^2$
30	$83^2 - 19^3$
40	$4^4 - 6^3 = 16^2 - 6^3$
50	?
60	$4^3 - 2^2$
80	$12^2 - 4^3$
100	$15^2 - 5^3 = 7^3 - 3^5$
200	$6^3 - 2^4$
500	$25^2 - 5^3$
600	$40^2 - 10^3 = 10^3 - 20^2$

It is interesting to note that Pierre de Fermat showed that 26 is the only number that is between a square and a cubic power: $5^2 = 25 < 26 < 27 = 3^3$. In other words, $x^3 - y^2 = 2$ has only one solution: $x = 3$ and $y = 5$.

POLIGNAC'S IDEA THAT LED TO MISTAKEN WORK

Another conjecture that has led to many mistaken "proofs" and along the way opened some new areas of mathematical research is the twin-prime conjecture. It was first stated in 1849 by Alphonse de Polignac (1817–1890). The conjecture is that there are an infinite number of prime twins; these are prime numbers that differ by 2, such as (3, 5), (5, 7), and (11, 13). With the exception of the first such prime twin, (3, 5), all the others seem to have a multiple of 6 between them. The table below shows a few of these prime twins.

n	$6n-1$	$6n+1$	Multiple of 6 between Prime Twins
1	5	7	6
2	11	13	12
3	17	19	18
5	29	31	30
7	41	43	42
10	59	61	60
12	71	73	72

Polignac actually stated that for any positive even number n, there are infinitely many prime intervals of size n. That is, there are infinitely many cases of two consecutive prime numbers with a difference of n. In the case of the prime twins, $n = 2$.

On December 25, 2011, the computer site www.Primegrid.com announced that it had found the to-date largest prime twin. It is $3756801695685 \cdot 2^{666669} \pm 1$. The numbers of this prime twin have 200,700 decimal digits.

However, the question as to whether there are an infinite number of prime twins still plagues mathematicians. Many mathematicians have produced "proofs" of this conjecture, only to find that others have found mistakes in the presented "proofs." Yet with all of these attempts, our understanding of mathematics improves.

A STILL-UNANSWERED QUESTION

Prime numbers are a source of fascination to mathematicians. For example, there is the question as to whether there exists an infinite number of prime numbers within many sequences of numbers. Questions such as this often lead mathematicians to make mistakes as they present their solutions—only to be thwarted as others find the faults in their work. This is particularly true of the Fibonacci numbers[14]: 1, 1, 2, 3, 5, 8, 13, 21, 34, 55, 89, . . . This series stems from a problem on the regeneration of rabbits in Leonardo of Pisa's (today, known as Fibonacci, ca. 1175–1240) book *Liber Abaci*, written in 1202. The sequence is recursive in that after the first two 1s, the succeeding number is always the sum of the two preceding numbers. The question as to whether the Fibonacci sequence has a finite or infinite number of prime numbers is still unanswered.

GOLDBACH'S CONJECTURES (AS YET UNSUBSTANTIATED) HAVE LED TO MANY MISTAKEN ATTEMPTS

German mathematician Christian Goldbach (1690–1764), in a June 7, 1742, letter to Leonhard Euler, posed the following statement, which to this day has yet to be solved:

Every odd number greater than 5 is the sum of three primes.

This is Goldbach's second conjecture or the weak Goldbach conjecture.

Euler strengthened this conjecture with the following—today known as Goldbach's first conjecture or the strong Goldbach conjecture:

Every even number greater than 2 can be expressed as the sum of two prime numbers.

You might want to begin with the following list of even numbers and their respective sum of prime numbers, and then continue it to convince yourself that it goes on—apparently—indefinitely.

Even Numbers Greater Than 2	Sum of Two Prime Numbers
4	2 + 2
6	3 + 3
8	3 + 5
10	3 + 7
12	5 + 7
14	7 + 7
16	5 + 11
18	7 + 11

20	7 + 13
. . .	. . .
48	19 + 29
. . .	. . .
100	3 + 97

Again, there have been substantial attempts at proving this conjecture by famous mathematicians: the German mathematician Georg Cantor (1845–1918) showed that the conjecture was true for all even numbers up to 1,000. It was then shown in 1940 to be true for all even numbers up to 100,000. By 1964, with the aid of a computer, it was extended to 33,000,000; in 1965, this was extended to 100,000,000; and then in 1980, to 200,000,000. Then, in 1998, the German mathematician Jörg Richstein showed that Goldbach's conjecture was true for all even numbers up to 400 trillion.

In April 2012, Tomás Oliveira e Silva extended this conjecture to $4 \cdot 10^{18}$. Prize money of $1,000,000 has been offered for a proof of this conjecture. To date, this has not been claimed.[15] Along the way, there were mistakes made when mathematicians tried to show this is true for all even numbers. But this still remains unsolved.

Unlike his other conjectures that seem to be correct, but have never been proved to be so, Goldbach made a further conjecture that was finally proved to be mistaken. In a letter to Euler on November 18, 1852, he stated that every odd number greater than 3 can be written as the sum of an odd number and twice a square number. Here are a few examples of his conjecture.

Odd Number	**Sum of a Prime Number and the Double of a Square Number** (Goldbach Considered 1 a Prime Number)
3	$= 1 + 2 \cdot 1^2$
5	$= 3 + 2 \cdot 1^2$
7	$= 5 + 2 \cdot 1^2$
9	$= 7 + 2 \cdot 1^2 = 1 + 2 \cdot 2^2$

11	$= 3 + 2 \cdot 2^2$
13	$= 5 + 2 \cdot 2^2 = 11 + 2 \cdot 1^2$
15	$= 7 + 2 \cdot 2^2 = 13 + 2 \cdot 1^2$
17	$= 17 + 2 \cdot 0^2$
19	$= 11 + 2 \cdot 2^2 = 17 + 2 \cdot 1^2$
21	$= 13 + 2 \cdot 2^2 = 19 + 2 \cdot 1^2$

Euler replied on December 16, 1752, stating that he checked the first 1,000 odd numbers and found it to hold true. On April 3, 1753, Euler wrote Goldbach again, this time saying that he showed it held true for the first 2,500 odd numbers. However, in 1856, the German mathematician Moritz Stern (1807–1894) found that Goldbach's conjecture did not hold true for the numbers 5,777 and 5,993, thereby rendering the conjecture a mistake. If one eliminates 0 as a possible square number, the Stern inspection of the first 9,000 odd numbers reveals the following that also cannot fit the Goldbach conjecture: 17, 137, 227, 977, 1187, and 1493. To date, no other counterexamples have been found. Through Goldbach's recreational view of mathematics—despite mistakes—he stimulated much research in number theory.

LOTHAR COLLATZ'S CONJECTURE GENERATES MISTAKES

In 1937, the German mathematician Lothar Collatz (1910–1990) posed a problem, known as the $3n+1$ problem (or Hasse's algorithm or the Ulam problem). Solutions to this problem have been attempted by numerous mathematicians without success. It, too, has experienced many mistaken attempts.

The problem is to prove that this conjecture is true.

Beginning with any *arbitrarily* selected number, do the following:

If the number is *odd*, then multiply by 3 and add 1.

If the number is *even*, then divide by 2.

Continue this process until you get into a repeating loop.

Regardless of the number you select, after continued repetitions of this process, one will always end up with the number 1.

Suppose we begin with the number 7; following this rule, we would get the following sequence of numbers: 7, 22, 11, 34, 17, 52, 26, 13, 40, 20, 10, 5, 16, 8, **4, 2, 1, 4, 2, 1, . . .**

Most recently on June 1, 2008,[16] using computers, this $3n + 1$ problem has been shown to be true for the numbers up to $18 \cdot 2^{58} \approx 5.188146770 \cdot 10^{18}$—that means, for more than 5 quintillion it is proved true.[17] However, it has still not been proved true for *all* cases.

MISTAKE-FREE EFFORTS PERSIST WITH LEGENDRE'S CONJECTURE

Mistakes have also been made in an attempt to solve the Legendre conjecture, made by Adrien-Marie Legendre, which states that for all natural numbers n, between n^2 and $(n + 1)^2$ there exists at least one prime number. The table below shows some of the first cases of this relationship.

n	1	2	3	4	5	6	7	8	9	10	11
n^2	1	4	9	16	25	36	49	64	81	100	121
$(n+1)^2$	4	9	16	25	36	49	64	81	100	121	144
p	2, ...	5, ...	11, ...	17, ...	29, ...	37, ...	53, ...	67, ...	83, ...	101, ...	127, ...

n	12	13	14	15	16	17	18	19	20
n^2	144	169	196	225	256	289	324	361	400
$(n+1)^2$	169	196	225	256	289	324	361	400	441
p	149, ...	173, ...	197, ...	227, ...	257, ...	293, ...	331, ...	367, ...	401, ...

The smallest primes between n^2 and $(n + 1)^2$ for $n = 1, 2, 3, \ldots$ are 2, 5, 11, 17, 29, 37, 53, 67, 83, . . .

The numbers of primes between n^2 and $(n + 1)^2$ for $n = 1, 2, 3, \ldots$ are given by 2, 2, 2, 3, 2, 4, 3, 4, . . .

For $n = 10$, there are five primes between 10^2 and 11^2, namely, 101, 103, 107, 109, and 113.

For $n = 1{,}000$, there are 152 primes between $(10^3)^2 = 10^6$ and $(10^3 + 1)^2 = 1{,}002{,}001$, namely:

1,000,003; 1,000,033; 1,000,037; 1,000,039; 1,000,081; 1,000,099; 1,000,117; 1,000,121; 1,000,133; 1,000,151; 1,000,159; 1,000,171; 1,000,183; 1,000,187; 1,000,193; 1,000,199; 1,000,211; 1,000,213; 1,000,231; 1,000,249; 1,000,253; 1,000,273; 1,000,289; 1,000,291; 1,000,303; 1,000,313; 1,000,333; 1,000,357; 1,000,367; 1,000,381; 1,000,393; 1,000,397; 1,000,403; 1,000,409; 1,000,423; 1,000,427; 1,000,429; 1,000,453; 1,000,457; 1,000,507; 1,000,537; 1,000,541; 1,000,547; 1,000,577; 1,000,579; 1,000,589; 1,000,609; 1,000,619; 1,000,621; 1,000,639; 1,000,651; 1,000,667; 1,000,669; 1,000,679; 1,000,691; 1,000,697; 1,000,721; 1,000,723; 1,000,763; 1,000,777; 1,000,793; 1,000,829; 1,000,847; 1,000,849; 1,000,859; 1,000,861; 1,000,889; 1,000,907; 1,000,919; 1,000,921; 1,000,931; 1,000,969; 1,000,973; 1,000,981; 1,000,999; 1,001,003; 1,001,017; 1,001,023; 1,001,027; 1,001,041; 1,001,069; 1,001,081; 1,001,087; 1,001,089; 1,001,093; 1,001,107; 1,001,123; 1,001,153; 1,001,159; 1,001,173; 1,001,177; 1,001,191; 1,001,197; 1,001,219; 1,001,237; 1,001,267; 1,001,279; 1,001,291; 1,001,303; 1,001,311; 1,001,321; 1,001,323; 1,001,327; 1,001,347; 1,001,353; 1,001,369; 1,001,381; 1,001,387; 1,001,389; 1,001,401; 1,001,411; 1,001,431; 1,001,447; 1,001,459; 1,001,467; 1,001,491; 1,001,501; 1,001,527; 1,001,531; 1,001,549; 1,001,551; 1,001,563; 1,001,569; 1,001,587; 1,001,593; 1,001,621; 1,001,629; 1,001,639; 1,001,659; 1,001,669; 1,001,683; 1,001,687; 1,001,713; 1,001,723; 1,001,743; 1,001,783; 1,001,797; 1,001,801; 1,001,807; 1,001,809; 1,001,821; 1,001,831; 1,001,839; 1,001,911; 1,001,933; 1,001,941; 1,001,947; 1,001,953; 1,001,977; 1,001,981; 1,001,983; 1,001,989.

Yet, to date, there has been no solution to this conjecture for all numbers.

However, this conjecture—although still unsolved—has been extended by Martin Aigner (1942–) and Günter Ziegler (1963–),[18] who have been in search of a mistake-free solution to a question raised by Ludwig von Oppermann (1882), which asks if there exist two primes between any two consecutive square numbers. And so it goes on with such conjectures that motivate mathematicians to find mistake-free solutions.

THE MISTAKES INVOLVING MERSENNE PRIMES

We recall that a prime number is a number greater than 1 whose only divisors are 1 and the number itself. There was a lot of intrigue regarding a formula for generating prime numbers. It should be clear that any number of the form $2^n - 1$ cannot be a prime number if n is not a prime number. Many early mathematicians thought that any number that can be written in the form $2^n - 1$ is a prime for all prime values of n.

k	$2^k - 1$	k Prime Number	$2^k - 1$ Prime Number	Factorization of $2^k - 1$
0	0	no	no	–
1	1	no	no	–
2	3	yes	yes	–
3	7	**yes**	**yes**	–
4	15	no	no	$3 \cdot 5$
5	31	**yes**	**yes**	–
6	63	no	no	$3^2 \cdot 7$
7	127	**yes**	**yes**	–
8	255	no	no	$3 \cdot 5 \cdot 17$
9	511	no	no	$7 \cdot 73$
10	1023	no	no	$3 \cdot 11 \cdot 31$

This was further supported in the mid-fifteenth century, when it was determined that $2^{13} - 1 = 8{,}191$, a prime number. This mistake in this conjecture was first discovered in 1536 by the German mathematician Ulrich Rieger (in Latin: Hudalricus Regius), who showed that $2^{11} - 1 = 2{,}047 = 23 \cdot 89$, and is, therefore, not a prime—even though $n = 11$ is a prime.

In 1603, Pietro Cataldi (1548–1626) showed that $2^{17} - 1$ and $2^{19} - 1$ were also primes and then went on, "pushing his luck" by mistakenly stating that $2^n - 1$ was also prime for $n = 23, 29, 31$, and 37. These mistakes were corrected by some famous mathematicians. In 1640, Fermat showed that Cataldi was wrong for the values $n = 23$ and $n = 37$; and in 1738, Euler showed that Cataldi was wrong in his assumption that $n = 29$ works, but he supported Cataldi's asumption of $n = 31$ as being correct.

This topic generated further mistakes. The French monk Marin Mersenne (1588–1648), after whom prime numbers of the form $2^n - 1$ are named, published a book in 1644, *Cognitata Physica—Mathematica.* In the preface he stated that numbers in the form $2^n - 1$ were prime for values of $n = 2, 3, 5, 7, 13, 17, 19, 31, 67, 127$, and 257, and were nonprime for all other numbers $n < 257$. This statement was a mistake! Although he claimed to have checked this conjecture, even his contemporaries did not really believe this claim.

It took mathematicians over one hundred years to begin to explore this claim. Again in 1750, Euler verified that $2^{31} - 1$ is prime; and in 1876, the French mathematician Édouard Lucas (1842–1891) showed that $2^{127} - 1$ is also a prime. For the next seventy years, this was known to be the largest prime number.

In 1883, the Russian mathematician Ivan M. Pervushin (1827–1900) showed that $2^{61} - 1 = 2{,}305{,}843{,}009{,}213{,}693{,}951$ is a prime number, which showed a mistake in Mersenne's conjecture, since he missed the number 61 on his list of possible values for *n.* To further highlight mistakes in Mersenne's conjecture, R. E. Powers showed in 1911 that 89 was an acceptable value for *n*, and in 1914 he showed that 107 was also an acceptable value for *n.* However, another mistake in Mersenne's conjecture was that $n = 67$ generated a prime number. We now see that $2^{67} - 1 = 147{,}573{,}952{,}589{,}676{,}412{,}927 = 193{,}707{,}721 \cdot 761{,}838{,}257{,}287$, and is, therefore, *not* a prime.

The list of acceptable values for *n* was finally checked up to 258 in 1947. They were $n = 2, 3, 5, 7, 13, 17, 19, 31, 61, 89, 107$, and 127.

Following is a list of these Mersenne numbers:

k	$2^k - 1$
2	3
3	7
5	31
7	127
13	8,191

17	131,071
19	524,287
31	2,147,483,647
61	2,305,843,009,213,693,951
89	618,970,019,642,690,137,449,562,111
107	162,259,276,829,213,363,391,578,010,288,127
127	170,141,183,460,469,231,731,687,303,715,884,105,727

Today we call a prime of the form $2^n - 1$ a Mersenne prime. In 1952, R. M. Robinson enlarged the list of Mersenne primes to include $n = 521, 607, 1{,}229, 2{,}203$, and 2,281. Today we have identified almost fifty Mersenne primes, of which the largest is $2^{43,112,609} - 1$ and has 12,978,189 decimal places!

To give some more significance to the Mersenne primes, we can look at their relationship to perfect numbers—that is, numbers where the sum of all of the proper divisors of the number (i.e., all the divisors except the number itself) is equal to the number itself. The smallest perfect number is 6, since $6 = 1 + 2 + 3$, which is the sum of all its divisors excluding the number 6 itself. The next larger perfect number is 28, since again $28 = 1 + 2 + 4 + 7 + 14$.

Some other perfect numbers are 496 and 8,128. Consider these numbers factored as:

$$6 = 2 \cdot 3$$
$$28 = 4 \cdot 7$$
$$496 = 16 \cdot 31$$
$$8{,}128 = 64 \cdot 127.$$

These numbers are in the form $2^{n-1}(2^n - 1)$, for $n = 2, 3, 5$, and 7. It was Euclid (ca. 365–ca. 310/290 BCE) who came up with a theorem to generalize a procedure to find a perfect number. He said that for an integer, n, if $2^n - 1$ is a prime number, then $2^{n-1}(2^n-1)$ is a perfect number.

Surely, we notice that one of the factors is a Mersenne prime $2^n - 1$. We can thus state that p is an even perfect number if and only if it has the form $2^{n-1}(2^n - 1)$ and $2^n - 1$ is a Mersenne prime number. By the way, if $2^n - 1$ is prime, then also n is a prime.

Before we leave perfect numbers, notice that they will all end in either a 6 or an 8.

After correcting the historical mistakes, Mersenne primes were celebrated by the United States Postal Service, when in 1963 the Canadian mathematician Donald B. Gillies (1929–1975) discovered[19] what was then the largest Mersenne prime. This prime was featured in the postage meter at the University of Illinois (see figure 1.14).

Figure 1.14. US postage meter stamp celebrating a Mersenne prime discovery.

It has still not been determined if there are an infinite number of perfect numbers. Additionally, we do not know whether or not there are any odd perfect numbers, but none has yet been found. We do know that if there are any such perfect numbers, they will have to be larger than 10^{1500} and have at least eight different prime number divisors (or eleven different prime number divisors, if the number is not divisible by three).[20]

PIERRE FERMAT'S BIG MISTAKE

One of the leading mathematicians of all time was the French mathematician Pierre de Fermat, who despite his fame was also prone to making a mistake. In a letter to the mathematician Blaise Pascal in 1654, he mentioned that although he had not yet developed a proof, he believed that all numbers of the form $F_m = 2^{2^m} + 1$, where m is a natural number, are prime numbers. Today these numbers are still called Fermat numbers. As you might expect by now, Fermat was wrong! Let's take a look at the first few such numbers.

$$F_0 = 2^{2^0} + 1 = 2^1 + 1 = 3$$

$$F_1 = 2^{2^1} + 1 = 2^2 + 1 = 5$$

$$F_2 = 2^{2^2} + 1 = 2^4 + 1 = 17$$

$$F_3 = 2^{2^3} + 1 = 2^8 + 1 = 257$$

$$F_4 = 2^{2^4} + 1 = 2^{16} + 1 = 65{,}537$$

However, when $m = 5$, we find that Fermat made a mistake, because F_5 is not a prime number. In 1732, Euler showed that $F_5 = 2^{2^5} + 1 = 2^{32} + 1 = 4{,}294{,}967{,}297$, which factors as $641 \cdot 6{,}700{,}417$, and is, therefore, not a prime number.

Fermat's error takes yet another step. In 1880, the French mathematician Fortune Landry (1799–?) at age eighty-two showed that the Fermat number $F_6 = 2^{2^6} + 1 = 2^{64} + 1$ is also not prime (also shown separately by H. Le Lasseur in 1880), since $F_6 = 18{,}446{,}744{,}073{,}709{,}551{,}617 = 274{,}177 \cdot 67{,}280{,}421{,}310{,}721$.

As history would have it, this factorization was already mentioned in 1855 in a letter from Thomas Clausen (1801–1885) to Carl Friedrich Gauss, yet it was not known widely at the time. So we must assume that Landry's work was independent of this information. Clausen did prove that 67,280,421,310,721 is a prime number.[21] To further "embarrass" Fermat, in 1975 Michael A. Morrison and John Brillhart published a paper that showed that F_7 is also not a prime, since it factors as:

$$F_7 = 340282366920938463463374607431768211457$$
$$= 59649589127497217 \cdot 5704689200685129054721.$$

Between 1980 and 1995, the Fermat numbers that were shown to be factorable are: F_8, F_9, F_{10}, and F_{11}. Others that are factorable are $F_{12} - F_{32}$. To date, the smallest Fermat number not yet determined to be factorable or prime is F_{33}. Actually, F_4 is the largest known prime Fermat number.

It is interesting to note that in 1796, Gauss showed that a regular n-sided polygon can be constructed with straightedge and compass if the odd prime factors of n are distinct Fermat primes. Gauss conjectured that this was also a necessary condition, but never proved it. However, the French mathematician Pierre Wantzel (1814–1848) proved it in 1837.

One of Gauss's most treasured discoveries was that a polygon of seventeen sides could be so constructed. He asked that it be engraved on his tombstone, and it was. The impossible regular n-sided polygon constructions are those where $n = 7, 11, 13, 19, 23, \ldots$, and for powers of primes, such as 9, 25, 27, . . . , and naturally for multiples of these numbers.

THE MISTAKEN CONJECTURE OF ALPHONSE DE POLIGNAC

The French mathematician Alphonse de Polignac (1817–1890) stated that "every odd number greater than 1 can be expressed as the sum of a power of 2 and a prime number."[22]

If we inspect the first few cases, we find that this appears to be a true statement. However, as you will see from the following list, it holds true for the odd numbers from 3 through 125 and then is *not* true for 127, after which it continues to hold true again for a while.

Odd Number	Sum of a Power of 2 and a Prime Number
3	$= 2^0 + 2$
5	$= 2^1 + 3$
7	$= 2^2 + 3$
9	$= 2^2 + 5$
11	$= 2^3 + 3$
13	$= 2^3 + 5$
15	$= 2^3 + 7$
17	$= 2^2 + 13$
19	$= 2^4 + 3$
. . .	. . .
51	$= 2^5 + 19$
. . .	. . .

125	$= 2^6 + 61$
127	= ?
129	$= 2^5 + 97$
131	$= 2^7 + 3$

Perhaps you can find the next number that fails de Polignac's conjecture. Indeed, the next numbers that fail de Polignac's conjecture are 149, 251, 331, 337, 373, and 509, while another counterexample is 877.

It is fairly easy to check for any one of these. It might seem that 127 is a "tricky" case, since it is a sum of powers of 2, so we'll check 149 instead.

Since $2^8 = 256 > 149$, we simply need to verify that $149 - 2^k$ is not prime for every value of $k = 1$ to 7. Just check:

$$
\begin{array}{ll}
149 - 2^0 = 149 - 1 = 148 & \text{(divisible by 2)} \\
149 - 2^1 = 149 - 2 = 147 & \text{(divisible by 3)} \\
149 - 2^2 = 149 - 4 = 145 & \text{(divisible by 5)} \\
149 - 2^3 = 149 - 8 = 141 & \text{(divisible by 3)} \\
149 - 2^4 = 149 - 16 = 133 & \text{(divisible by 7)} \\
149 - 2^5 = 149 - 32 = 117 & \text{(divisible by 3)} \\
149 - 2^6 = 149 - 64 = 85 & \text{(divisible by 5)} \\
149 - 2^7 = 149 - 128 = 21 & \text{(divisible by 3)}
\end{array}
$$

You can do the same for other counterexamples to see that they also fail to have any kind of decomposition into 2^k + prime, since we can check all possible 2^k and see that the difference is never prime.

In 1848, Polignac further conjectured that every odd number greater than 1 and smaller than 3,000,000 (without the number 959) can be expressed as the sum of a power of 2 and a prime number.[23] It was not until 1960 that it was proved[24] that there are infinitely many odd numbers that exhibit Polignac's mistake; for example, the number 2,999,999 cannot be expressed as the sum of a power of 2 and a prime number.

By the way, in 1849, Alphonse de Polignac proposed another conjecture that to date has not been proved or disproved. It is as follows:

There are infinitely many cases of two consecutive prime numbers with difference of some *even* number n.

For example, suppose we let $n = 2$. There are consecutive prime number pairs whose difference is 2, such as (3, 5), (11, 13), (17, 19), and so on. Again, we still have not established if this conjecture is true or false.

LEONHARD EULER'S MISTAKEN CONJECTURE

Mistakes by famous mathematicians have often led to many new—and sometimes unrelated—discoveries. Yet, for one of the most prolific mathematicians, the Swiss mathematician Leonhard Euler, to make a conjecture that was mistaken is quite astonishing; yet, it has motivated mathematicians for centuries. We are all familiar with the Pythagorean theorem, where we know that there are integer solutions to $a^2 + b^2 = c^2$. Euler had proved that $a^3 + b^3 = c^3$ has no integer solution. However, through the fine efforts in 1994 of the British mathematician Andrew Wiles, we know that the equation $a^n + b^n = c^n$ has no integer solutions when $n > 2$.

Building upon his finding, Euler conjectured that none of the following equations, and those analogously formed, have integer solutions:

$$a^3 + b^3 = c^3$$
$$a^4 + b^4 + c^4 = d^4$$
$$a^5 + b^5 + c^5 + d^5 = e^5, \text{ and so on.}$$

Well, as fate would have it, Euler was proved wrong! In 1966, Leon J. Lander and Thomas R. Parkin found a solution for $n = 5$:[25]

$$27^5 + 84^5 + 110^5 + 133^5 = 61{,}917{,}364{,}224 = 144^5.$$

Then, in 1988, Noam D. Elkies (1966–) found a solution for the case $n = 4$:[26]

$$2682440^4 + 15365639^4 + 18796760^4$$
$$= 180630077292169281088848499041 = 20615673^4.$$

Furthermore, Elkies showed that for $n = 4$ there are infinitely many solutions, the smallest of which was found in the same year by Roger Frye:

$$95800^4 + 217519^4 + 414560^4 = 31{,}858{,}749{,}840{,}007{,}945{,}920{,}321 = 422481^4.$$

These counterexamples exhibited the mistake in Euler's conjecture.

ANOTHER EULER MISTAKE

Before we besmirch one of the most revered mathematicians in history, we should say that his conjectures in many cases have held up and have led to many new directions in the study of mathematics. However, there have been times when he made a mistake. One such is his response to the problem that was posed in the court of Catherine the Great (1729–1796): Namely, how can six regiments, each consisting of six different officer titles (e.g., general, colonel, lieutenant colonel, major, captain, and lieutenant) be arranged in six rows and six columns so that no row or column has two of the same rank or regiment in it. Euler called this "the thirty-six-officers problem" in his *Recherches sur une Nouvelle Espece de Quarres Magiques* in 1782, and correctly conjectured that this was not possible. A proof of this conjecture did not come until 1900 by Gaston Tarry.[27]

The problem would be reduced to setting up a 6 × 6 square, where no repetitions of rank or regiment can be found in any row or column.

However, for a 4 × 4 square it is possible to arrange four suits with four players—as with playing cards, as shown in figure 1.15. Notice there is no repetition or any suit or face card in any row or column.

♥K	♣Q	♠J	♦A
♣J	♥A	♦K	♠Q
♠A	♦J	♥Q	♣K
♦Q	♠K	♣A	♥J

Figure 1.15.

Euler's mistake was that he further conjectured that it was impossible to set up a square arrangement of this type (with no repetitions) not only for a 2×2 square (which is trivial) and a 6×6 square as stated above, but for any square, where the number of rows (or columns) is in the form $4k+2$; that is, for 10×10 ($k=2$), or 14×14 ($k=3$) squares as well. It took centuries by some of the finest mathematicians to finally disprove Euler's conjecture.

As recently as 1922, it was believed that Euler's conjecture was proved correct.[28] However, in 1958, Raj Chandra Bose (1901–1987) and Sharadchandra Shankar Shrikhande (1917–) were able to refute this notion.[29] They were able to produce an appropriate 22×22 square—that is for the case where $4k + 2 = 22$, where $k = 5$. Their discovery was met with fanfare; so much so that the cover of the *Scientific American* (November 1959) exhibited this square. In the following year, Ernest Tilden Parker (1926–1991) constructed a 10×10 square.[30] Then, in 1960, R. C. Bose, S. S. Shrikhande, and E. T. Parker,[31] proved that the Euler conjecture is false for all $m = 4k + 2$ ($k \in \mathbf{N}$, $k > 1$).

Yet along the way, many new areas of mathematics were conceived. Thus Euler's partially correct conjecture proved to be of great value over the past few centuries.

THE EMBARRASSING MISTAKE BY LEGENDRE

The famous French mathematician Adrien-Marie Legendre, through Charles Davies's translation (and some reworking) of his 1794 French geometry book, gave American schools the high-school geometry course in the mid-nineteenth century. Yet he made a conjecture that was shown later to be not correct. He stated that there do not exist natural numbers p, q, r, and s such that

$$\left(\frac{p}{q}\right)^3 + \left(\frac{r}{s}\right)^3 = 6.$$

The British mathematician Henry Ernest Dudeney (1857–1930), who seemed to devote much time to recreational mathematics, discovered a counterexample to Legendre's conjecture and thereby rendered it a mistaken conjecture. He found that for $p = 17$, $q = 21$, $r = 37$, and $s = 21$, we get:

$$\left(\frac{17}{21}\right)^3 + \left(\frac{37}{21}\right)^3 = \frac{4913}{9261} + \frac{50653}{9261} = 6.$$

The mistake is identified!

THE UNEXPECTED MISTAKE BY NIKOLAI GRIGORIEVICH CHEBOTARYOV

Binomials $x^n - 1$ can be factored such that a factor of $x - 1$ can be extracted, as shown in the following table.

n	$x^n - 1$	Factorization
1	$x^1 - 1$	$x - 1$
2	$x^2 - 1$	$(x - 1) \cdot (x + 1)$
3	$x^3 - 1$	$(x - 1) \cdot (x^2 + x + 1)$
4	$x^4 - 1$	$(x - 1) \cdot (x + 1) \cdot (x^2 + 1)$
5	$x^5 - 1$	$(x - 1) \cdot (x^4 + x^3 + x^2 + x + 1)$
6	$x^6 - 1$	$(x - 1) \cdot (x + 1) \cdot (x^2 + x + 1) \cdot (x^2 - x + 1)$
7	$x^7 - 1$	$(x - 1) \cdot (x^6 + x^5 + x^4 + x^3 + x^2 + x + 1)$
8	$x^8 - 1$	$(x - 1) \cdot (x + 1) \cdot (x^2 + 1) \cdot (x^4 + 1)$
9	$x^9 - 1$	$(x - 1) \cdot (x^2 + x + 1) \cdot (x^6 + x^3 + x^2 + 1)$
10	$x^{10} - 1$	$(x - 1) \cdot (x + 1) \cdot (x^4 + x^3 + x^2 + x + 1) \cdot (x^4 - x^3 + x^2 - x + 1)$

Notice that all of the factors consist of constants and coefficients that are either +1 or –1. Taking this a bit further, we find that factoring for the case where $n = 20$, we get: $x^{20} - 1 = (x - 1) \cdot (x + 1) \cdot (x^2 + 1) \cdot (x^4 + x^3 + x^2 + x + 1) \cdot (x^4 - x^3 + x^2 - x + 1) \cdot (x^8 - x^6 + x^4 - x^2 + 1)$.

As n increases, the factorization becomes ever so much more complicated—that is, without using a computer algebra system (CAS). However, as n increases, we notice that the coefficient of x is always 1 or 0.

In 1938, the Soviet mathematician Nikolai Grigorievich Chebotaryov (1894–1947) stated—without proof—that this pattern will be true for all values of $n > 0$. This statement didn't last long before it was shown to be mistaken.

In 1941, another Soviet mathematician, Valentin Konstantinovich Ivanov (1908–1992), discovered a counterexample, specifically for $n = 105$, where we have[32]:

$$
\begin{aligned}
x^{105} - 1 = {} & (x-1)\cdot(x^2+x+1)\cdot(x^4+x^3+x^2+x+1) \\
& \cdot(x^6+x^5+x^4+x^3+x^2+x+1) \\
& \cdot(x^8-x^7+x^5-x^4+x^3-x+1) \\
& \cdot(x^{12}-x^{11}+x^9-x^8+x^6-x^4+x^3-x+1) \\
& \cdot(x^{24}-x^{23}+x^{19}-x^{18}+x^{17}-x^{16}+x^{14}-x^{13}+x^{12}-x^{11}+x^{10}-x^8+x^7-x^6 \\
& \quad -x^5-x+1) \\
& \cdot(x^{48}+x^{47}+x^{46}-x^{43}-x^{42}-\mathbf{2x^{41}}-x^{40}-x^{39}+x^{36}+x^{35}+x^{34}+x^{33}+x^{32}+x^{31} \\
& \quad -x^{28}-x^{26}-x^{24}-x^{22}-x^{20}+x^{17}+x^{16}+x^{15}+x^{14}+x^{13}+x^{12}-x^9-x^8-\mathbf{2x^7} \\
& \quad -x^6-x^5+x^2+x+1).
\end{aligned}
$$

Notice that with this factorization the coefficient –2 comes up twice (bold above): once as a coefficient for x^{41} and once for x^7. Just to take this a bit further—to emphasize the mistake of the conjecture—using a CAS we can see that the binomial $x^{2805} - 1$ uses (besides the +1 and –1) ,+2 and –2, +3 and –3, +4 and –4, +5 and –5, as well as +6 and –6, as in the sample term or the factorization: for instance $-6x^{707}$ and $+6x^{692}$.

HENRI POINCARÉ'S COSTLY MISTAKE

The French mathematician and physicist Henri Poincaré (1854–1912) was seen as one of the leading mathematicians of the 1890s through the beginning of twentieth century. In physics his contributions centered on optics, electricity, and quantum theory. He was also concerned with thermodynamics and the theory of relativity, of which he was one of the founders. King Oskar II of Sweden and Norway (1829–1907), in recognition of his sixtieth birthday, offered a prize of 2,500 Krona for the answers to four mathematical questions. The first of the four questions concerned the n-body problem. The background of this problem regarded the question: Would our solar system go on as it has forever, or would Earth distance itself from the sun in a spiral pattern? In other words, given n mass points, which we can imagine being the sun and the planets with their position and speed at a specific point in time, is it then possible to determine the movement of these bodies indefinitely into the future? For $n = 2$, Newton showed that such a system will remain in constant motion in an elliptical path with each body moving around their common mass center. Since it would have been too complicated to consider the planets of the solar system with their moons, Poincaré decided to consider a system with three planets. Newton was of the opinion that when $n > 2$, the computation would be so complicated that the human mind would not be able to tackle it. Solving this problem with three bodies was seen by Johannes Kepler and Nicholas Copernicus as one of the most difficult mathematical problems. Even Leonhard Euler and Joseph-Louis Lagrange (1736–1813) had tried without success to solve this problem. Portions of this problem deal with power series and geometric theorems related to differential equations. Poincaré was able to simplify this problem such that he was able to take some definitive steps toward its solution. He was also convinced that his approximations of the paths traveled would take him closer to the solution, and would not be affected by his rounding off the position of the three bodies' positions.

Poincaré's paper that examined the issue for the reduced 3-body problem consisted of 158 pages, which was extraordinarily long. Although it did not answer the original question completely, it was nevertheless accepted, and he won the prize. After the prizes were awarded, it was discovered that there was a slight change in the positions of the planets, which resulted in a discrepancy in the path, and that brought a fault to his paper. Poincaré was distraught about the error, so much so that he approached the prize-awarding committee to return his prize. He had to acknowledge that even the slightest changes in his original conditions led to completely different paths. Poincaré subsequently wrote a follow-up paper because of his error that showed on further reflection that chaotic solutions are possible.

PRIZES FOR AVOIDING MISTAKES

Unsolved problems have a very important role in mathematics. Attempts to solve them oftentimes lead to very important findings of other sorts. An unsolved problem, one not yet solved by the world's most brilliant minds, tends to pique our interest by quietly asking us if we can solve it, especially when the problem itself is exceedingly easy to understand.

The German mathematician David Hilbert (1862–1943) announced his seminal list of twenty-three unsolved mathematical problems at the second International Congress of Mathematicians in Paris on August 8, 1900. Much of the mathematical research in the twentieth century has been influenced by this list of unsolved problems, as both successful and unsuccessful attempts at solutions have yielded a number of important discoveries along the way.

To commemorate this magnificent occasion and to provide a suitable launch for mathematics into a twenty-first century (one hundred years later), the newly formed Clay Mathematics Institute, located in Cambridge, Massachusetts, then devised its own list of problems that still lacked a solution and formally announced it at the Collège de France in Paris on May 24, 2000, in a lecture titled "The Importance of Mathematics." The founder and sponsor of the Clay Mathematics Institute, Landon T. Clay (1927–), a businessman who loved mathematics yet majored in English at Harvard University, believed mathematics research was underfunded and was interested in popularizing the subject. He offered $1,000,000 to anyone who could solve any of the listed—heretofore unsolved—mathematics problems. One of these "millennium problems" has to date been solved by the Russian mathematician Grigori Perelman (1966–), who in 2002 verified the Poincaré conjecture.

CARELESS MISTAKES BY THE FAMOUS!

There are simple mistakes in mathematics that are made carelessly by famous scientists. Unfortunately, they are sometimes logged in for eternity.

Take, for example, the error made by the famous scientist and Nobel laureate Enrico Fermi (1901–1954), known by his colleagues as "the Pope" since he rarely ever made a mistake. Well, suffice it to say, he did on occasion make mistakes, and sadly for him, one was photographed; and to compound this embarrassment, this photo from 1948 appeared on a United States postage stamp in 2001 to commemorate the one hundredth anniversary of his birth. The equation that he wrote on the chalkboard as shown in figure 1.16 has a mistake at the upper left of the stamp. This is a careless mistake by a marvelous scientist—Fermi mistakenly interchanged the symbols $\hbar$ and e.

34¢:
Wrong equation:

$$\alpha = \frac{\hbar^2}{e \cdot c}.$$

Correct equation:

$$\alpha = \frac{e^2}{\hbar \cdot c}.$$

α is the fine-structure constant;
e is the elementary charge;
$\hbar = h/2\pi$ is the reduced Planck constant; and
c is the speed of light in vacuum.

Figure 1.16.

CONCLUSION

As we close this discussion of mistakes made by brilliant mathematicians, we should not forget that one of the greatest modern minds, that of Albert Einstein (1879–1955), produced so many mistakes among his many discoveries that books have been written about them. One such book, *Einstein's Mistakes* by Hans C. Ohanian,[33] mentions that of Einstein's 180 published papers, forty of them had mistakes in them. These mistakes did not deter him from making his brilliant observations and were later corrected. In recent years, the notes of Johannes Kepler were found to contain mistakes of his planetary observations, but, again, Kepler overlooked these slight deviations and still made some of the most amazing discoveries regarding planetary motion and the properties of ellipses. Perhaps Einstein sums up this situation best with the famous quote often attributed to him: "He who has never made a mistake has never tried something new." Readers might wonder why we, here, are highlighting mistakes—particularly when they are not well known. Most researchers—scientists, mathematicians, and others who seek to break new ground—encounter mistakes, and oftentimes they can correct them. In other cases, as with Andrew Wiles, others identified these mistakes and the mistakes were corrected—at least we hope! Why, then, do we focus so little on these mistakes? Most books on science and mathematics do not mention the mistakes, only the correct results.

Yet, as we reflect on the history of mathematics, we notice that one of the leading mathematicians of all time, Carl Friedrich Gauss, seems never to have made any mistakes in any of his published works. He seemed to have been governed by the Latin adage *Pauca sed matura*, which means "few, but ripe." It was not until 1898, when his diary was discovered and analyzed, that we realize today that Gauss did not publish all of his discoveries.

Might we learn more about the various subjects if we were to be exposed to these famous mistakes? That is one of the goals of the remainder of this book.

CHAPTER 2

MISTAKES IN ARITHMETIC

As we embark on our journey through the many mistakes that occur in mathematics, it is reasonable that our first stop should be the area of arithmetic, since that is usually the initial contact that one has with mathematics. The mistakes here will range from counting errors to strange calculations and violations of logical thinking. Clearly some mistakes come from violating well-entrenched rules of mathematics—some of which are not as well known as they should be. Then there are mistakes that result from coming to conclusions too quickly and finding that they are incorrectly established. Let's begin the journey!

STUMBLING BLOCKS: COUNTING MISTAKES

A common mistake that students make, and that teachers think has long been corrected, can be seen with a typical subtraction problem. Consider the following:

Along the street, where the house numbers are marked consecutively from 22 through 57, the question is how many houses are there along the street.

Most students will simply subtract $57 - 22 = 35$, and answer 35 houses. This is clearly a wrong answer. Actually, $35 + 1 = 36$ is the right answer. To demonstrate the counting of houses, students could subtract 21 from both house numbers so the counting will begin with the house number 1, and the answer is then obvious.

There are analogous examples that shed further light on this rather common mistake. Consider a ten-story apartment building, where the first floor is the ground floor, and the staircases connecting floors all have the same number of stairs. How much longer is the walk to the tenth floor than to the fifth floor? The immediate answer typically is that it is twice as long. Unfortunately, this is a wrong answer. Since the first floor is the ground floor, there are nine flights to the tenth floor, but only four flights to the fifth floor. Therefore, it is $\frac{9}{4} = 2.25$ times as long to get to the tenth floor as to the fifth floor.

Analogous to this is a question about a clock that strikes five times at 5:00, taking five seconds (assuming there is no time taken by the "bong" of the clock). How long will it take this clock to strike ten at 10:00? The usual answer is ten seconds, but that is wrong! Because there are only four intervals during the five seconds when the clock strikes five times at 5:00, each interval has the duration of $\frac{5}{4}$ seconds. There are nine intervals when the clock strikes ten at 10:00. Therefore the time required for that is $9 \cdot \frac{5}{4} = 11\frac{1}{4}$ seconds.

There are countless other examples of this type that can be brought to the fore, and provide entertainment, because they are somewhat counterintuitive. We might call these common mistakes!

A MISTAKE OF NOT THINKING BEFORE CALCULATING

A book louse, an insect that likes to eat through book pages has set on a path through a three-volume set of books of identical size. The books are located on a shelf in their proper order from left to right. The book louse begins its boring journey at page 1 of volume 1 and through volume 2 until it gets to the last page of volume 3. If each of the books from cover to cover measures 5 cm, and each cover has a thickness of 1 mm, then what is the total length of the book louse's path (assuming it is a straight line path of minimal length).

The typical answer to this question is 14.8 cm, which is calculated from:

([Vol. 1 – 2 covers] + 1 cover) + (Vol. 2) + ([Vol. 3 – 2 covers] + 1 cover)
(4.8 cm + 0.1 cm) + (5 cm) + (4.8 cm + 0.1 cm).

This is wrong!

The correct answer is 5.2 cm, which comes from 0.1 cm + 5 cm + 0.1 cm.

If you look at figure 2.1, you will see that page 1 of volume 1 is at the right side of the book, and the last page of volume 3 is at the left side of volume 3.

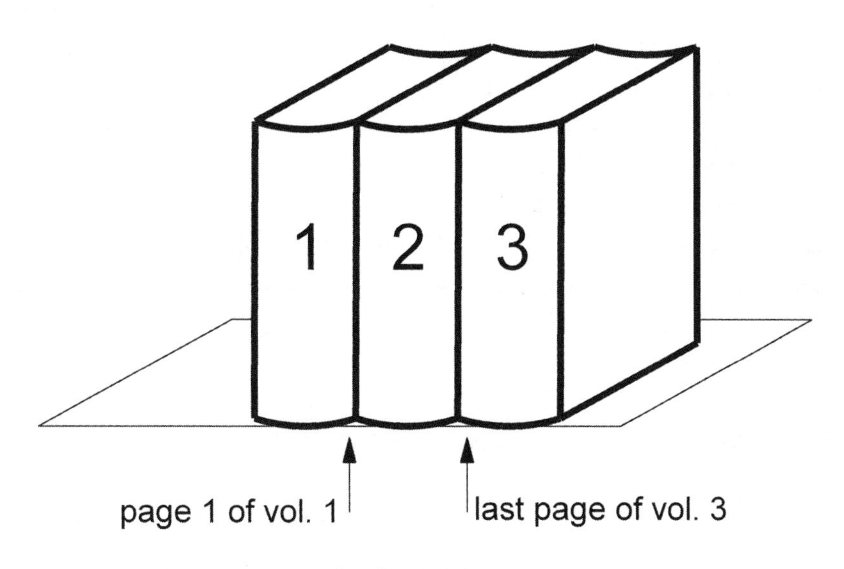

Figure 2.1.

STUMBLING BLOCKS: NUMBERING MISTAKES

There are oftentimes counting errors that occur and remain unnoticed for a long time. On January 1, 2000, the *New York Times* corrected an error that had been made more than one hundred years earlier. On February 6, 1898, an employee of this newspaper noticed that the current day's issue was number 14,499. And so he mistakenly gave the following day's issue number 15,000, instead of 14,500. It was not until Saturday, January 1, 2000, that this mistake was corrected. On that day, issue number 51,254 was published, while the previous day's issue was numbered 51,753. In case you're wondering, issue number 1 of the *New York Times* was published on September 18, 1851.

Some printing errors are not as easy to correct as that of the *New York Times*'s numbering system. We have been shown through the *Popeye* comics that spinach brings a person extra strength. This is largely a result of some misunderstandings—or perhaps mistakes—about the iron content in spinach. Spinach has approximately 3.5 mg of iron per 100 g of spinach; while in its cooked version, it contains about 2 mg, which turns out to be a lot less than bread, meat, or fish. This misunderstanding of spinach's iron value stems back to the 1930s, when there was a printing mistake. A decimal point was mistakenly moved one place to the right, which, of course, gives a value ten times that of the intended value. One might see this mistake as one where the value of iron in spinach was given ten times the correct amount, namely, 35 mg per 100 g of fresh spinach.[1] With *Popeye*'s reinforcement, many children grew up imagining spinach as an instant source of power.

ROUNDING OFF CAN CAUSE A MISTAKEN ANSWER

There are times when rounding off correctly leads to a wrong answer. Consider the following example: At an airport there are 963 stranded passengers. Busses are ordered to take these passengers on their way. Each bus can hold 59 passengers. The question is, how many buses are required to take these passengers on their way? Typically, a student would do the following calculation. $963 \div 59 \approx 16.32203389$. Since the number of buses must be a whole number, the student would round off the answer correctly to 16, since the 3 following the decimal point is less than 5. This answer clearly does not solve the problem. And although the calculation was correct, the problem was solved incorrectly. Obviously seventeen buses will be required,where the seventeenth bus will not be completely full. Here is an example of a mistaken answer despite the fact that correct calculations were made.

STUMBLING BLOCKS: THE NEMESIS OF THE ZERO

Let's inspect the following arithmetic: $\frac{0}{3}=0$, $\frac{0}{5}=0$; therefore, we might conclude that $\frac{0}{0}=0$, since the numerator is zero. $\frac{3}{3}=1$, $\frac{5}{5}=1$, therefore, we might conclude that $\frac{0}{0}=1$, since the numerator and the denominator are equal. Because of this dilemma, we call this fraction $\frac{0}{0}$ one of indeterminate form, and it, therefore, has no value.

Why does the zero play such a confusing role in arithmetic? We notice that in addition the zero plays an absolutely neutral role. This can be seen in the following example: $a + 0 = 0 + a = a$ for all (real) numbers a.

However, with subtraction we find the first stumbling block created by the zero. There is a difference if we subtract zero from a number, or a number from zero: $a - 0 = a$, whereas $0 - a = -a$ for all (real) numbers a.

The situation gets more complicated when we look at multiplication, and even more so when we look at division involving zero a bit later. In the case of multiplication, students see that zero seems to swallow up anything with which it is multiplied; in other words, zero times anything is zero: $a \cdot 0 = 0 \cdot a = 0$ for all (real) numbers a.

Another stumbling block for some is the confusion of the role of 0 and 1 in multiplication and addition. We know that when we multiply any number by 1, the number stays the same. Analogously, when we add zero to any number, that number also remains the same. As we can see from the following: $a \cdot 1 = 1 \cdot a = a$, $a + 0 = 0 + a = a$, for all (real) numbers a.

The situation gets more dramatic when we consider division involving zero. It makes a big difference whether the zero appears in the numerator or the denominator. If we divide zero by any number, we get zero: $0 \div a = \frac{0}{a} = 0$, for all (real) numbers a.

However, if we divide any number by zero, we end up with an inexplicable situation. Thus we say division by zero is not allowed, and that, many mathematics teachers instructed students, is "the eleventh commandment." Thou shall not divide by zero! Violation of this commandment will become clear in the following discussion. So brace yourself for some magnificent mistakes.

There are many ways in which we can "prove" that $1 = 2$. Now, one wonders, what does this mean? How can one prove that $1 = 2$? Clearly, there must be an error somewhere in the "proof." What's more, if this error is well hidden, then our "proof" becomes even more frustrating. Yet, knowing that this cannot be true, we search on, in the hope of locating this mistake. We shall begin with a simple mistake, one that can be seen from various points of view.

VIOLATING THE "ELEVENTH COMMANDMENT"—A MISTAKE!

We can "prove" that $1 = 2$ by using the forbidden division by zero. We know that if $5a = 5b$ then $a = b$. However, if we use the reasoning that since $1 \cdot 0 = 0$, and $2 \cdot 0 = 0$, we can conclude that $1 \cdot 0 = 2 \cdot 0$. Then, using the same reasoning as before, we can say that $1 = 2$, by dividing both sides by zero. Since such absurdities result from division by zero, it is forbidden usage in mathematics. Remember, the eleventh commandment: Thou shall not divide by zero.

A simple example of a violation of this rule can be seen with the following:

$$\begin{aligned}
&12 - 12 = 18 - 18 \\
&12 - 8 - 4 = 18 - 12 - 6 \\
&2 \cdot (6 - 4 - 2) = 3 \cdot (6 - 4 - 2) \\
&2 = 3
\end{aligned}$$

We can even create a comical situation by violating this important rule. That is, we can actually show through the following argument that no one is overweight. In other words, everybody's weight is correct for their body.

Follow this proof carefully. We shall begin by letting G = actual weight, g = ideal weight, W = overweight.

The obvious definition of our actual weight, $G = g + W$.

Subtract g from both sides of the equation to get: $G - g = W$.

Multiply both sides by $(G - g)$: $(G - g)^2 = W \cdot (G - g)$.

Do the indicated multiplication to get: $G^2 - 2G \cdot g + g^2 = G \cdot W - g \cdot W$.

Subtract g^2 from both sides of the equation to get:
$G^2 - 2G \cdot g = G \cdot W - g^2 - g \cdot W$.

Now subtract $G \cdot W$ from both sides of the equation, which gives us: $G^2 - 2G \cdot g - G \cdot W = -g^2 - g \cdot W$.

We will now add $G \cdot g$ to both sides of the equation to get: $G^2 - G \cdot g - G \cdot W = G \cdot g - g^2 - g \cdot W$.

Simplifying further gives us: $G \cdot (G - g - W) = g \cdot (G - g - W)$.

Dividing both sides by $(G - g - W)$ leaves us with: $G = g$.

Which essentially states that our actual weight is our ideal weight. We then ask ourselves, where does the mistake lurk? Well, when we recall that $G = g + W$, then $G - g - W = 0$. Then take note of the step that led us to the absurd conclusion above.

ADDING FRACTIONS: WRONG METHOD, RIGHT SUM!

Suppose we want to add the two fractions: $\frac{9}{3} + \frac{-16}{4}$. By adding the numerators and the denominators—clearly, a wrong procedure—we have $\frac{9}{3} + \frac{-16}{4} = \frac{9-16}{3+4}$, which surprisingly gives us a correct answer, $\frac{9-16}{3+4} = \frac{-7}{7} = -1$. Remember, of course, that the procedure was certainly wrong.

The correct procedure, $\frac{9}{3} + \frac{-16}{4} = 3 - 4 = -1$, as you can see, gives us the correct answer, which is the same as the one before. Thus, a correct answer does not support an incorrect procedure!

We can look at another example of this faulty procedure, which, curiously, also leads to a correct answer, as follows: $\frac{-5}{-1} + \frac{20}{2} = \frac{-5+20}{-1+2} = \frac{15}{1} = 15$. Done correctly, it would look like this $\frac{-5}{-1} + \frac{20}{2} = 5 + 10 = 15$, which is, amazingly, the same answer. Don't be fooled to believe that this procedure is a simplified way of adding fractions, even though it leads to a correct answer in this particular case. It is still a mistaken procedure.

A CRAZY (MISTAKEN) PROCEDURE TO MULTIPLY FRACTIONS LEADS TO THE CORRECT ANSWER

Here we are given two binomials consisting of fractions, which we then multiply in a rather strange way:

$$\left(\frac{2}{3}-\frac{5}{4}\right)\cdot\left(\frac{1}{3}+\frac{5}{8}\right)=\frac{2}{3}\cdot\frac{1}{3}-\frac{5}{4}\cdot\frac{5}{8}$$

This erroneous procedure has us multiplying the first two members of each parenthetical expression and the second two members of each parenthetical expression. Doing the indicated arithmetic, we get the following correct answer:

$$\frac{2}{3}\cdot\frac{1}{3}-\frac{5}{4}\cdot\frac{5}{8}=\frac{2}{9}-\frac{25}{32}=-\frac{161}{288}=-0.55902\overline{7}\,.$$

Comparing this to the correct procedure—just to make sure that, in fact, this answer is truly correct:

$$\left(\frac{2}{3}-\frac{5}{4}\right)\cdot\left(\frac{1}{3}+\frac{5}{8}\right)=-\frac{7}{12}\cdot\frac{23}{24}=-\frac{161}{288}=-0.55902\overline{7}\,.$$

Once again, it was a matter of luck that the correct answer was obtained with a completely mistaken procedure. This should teach us that if something "works" for one case, it is not immediately generalizable.

SOME MISTAKES BASED ON HASTILY JUMPING TO CONCLUSIONS

One can easily establish that the following numbers are all prime numbers:

31;
331;
3,331;
33,331;
333,331;
3,333,331;
33,333,331.

First we should note that these numbers are of the form $\frac{10^n-7}{3}$ (n = 2, 3, 4, . . . , 8). (You may also recall that the number 31 is a Mersenne prime number, since $31 = 2^5 - 1$.)

One could easily draw the conclusion that all members of the form 333 . . . 3,331 are prime numbers. However, this would be a mistake, and such mistakes have often been made in the history of mathematics; and yet these mistakes have then been the springboard for much further research and relevant findings. In the chart below, notice what happens when *n* takes on values greater than 8.

n	$\frac{10^n-7}{3}$		**Prime Factorization**
9	333,333,331	=	17 · 19,607,843
10	3,333,333,331	=	673 · 4,952,947
11	33,333,333,331	=	307 · 108,577,633
12	333,333,333,331	=	19 · 83 · 211,371,803
13	3,333,333,333,331	=	523 · 3,049 · 2,090,353
14	33,333,333,333,331	=	607 · 1,511 · 1,997·18,199
15	333,333,333,333,331	=	181 · 1,841,620,626,151
16	3,333,333,333,333,331	=	199 · 16,750,418,760,469
17	33,333,333,333,333,331	=	31 · 1,499 · 717,324,094,199

The recurrence of the factor 31 in the last number shows that no sequence of this type can consist only of primes, because every prime in the sequence will periodically divide further numbers. In the case above, we notice that the number 31 divides every fifteenth number of the form 333 . . . 33,331. And we should take note that the number 331 divides every 110th number of the form 333 . . . 3,331.

This time we will begin with a different pattern of numbers:

91; 9,901; 999,001; 99,990,001; 9,999,900,001; 999,999,000,001; . . .

When we inspect these numbers, we noticed that there is a pattern emerging, namely, beginning with the second number, every other one is a prime number.

n	$10^{2n} - 10^n + 1$			Prime or Composite
1	$10^2 - 10^1 + 1$	=	91	$= 7 \cdot 13$
2	$10^4 - 10^2 + 1$	=	9,**90**1	Prime
3	$10^6 - 10^3 + 1$	=	99**9,0**01	$= 19 \cdot 52{,}579$
4	$10^8 - 10^4 + 1$	=	99,9**90**,001	Prime
5	$10^{10} - 10^5 + 1$	=	9,999,**90**0,001	$= 7 \cdot 13 \cdot 211 \cdot 241 \cdot 2{,}161$
6	$10^{12} - 10^6 + 1$	=	999,99**9,0**00,001	Prime
7	$10^{14} - 10^7 + 1$	=	99,999,9**90**,000,001	$= 7^2 \cdot 13 \cdot 127 \cdot 2{,}689 \cdot 459{,}691$
8	$10^{16} - 10^8 + 1$	=	9,999,999,**90**0,000,001	Prime
9	$10^{18} - 10^9 + 1$	=	999,999,99**9,0**00,000,001	$= 70{,}541{,}929 \cdot 14{,}175{,}966{,}169$

Much to our disappointment, making a generalization here would be a mistake. For, as we might expect the tenth number to be a prime number, it is not.

For $n = 10$, we have: $10^{20} - 10^{10} + 1 = 99{,}999{,}999{,}990{,}000{,}000{,}001$ $= 61 \cdot 9{,}901 \cdot 4{,}188{,}901 \cdot 39{,}526{,}741$.

By the way, for $n = 12$, a prime number is also not generated.

Let's depart now from prime numbers, since they don't seem to want to follow a pattern anyway.

We will look at the product of two numbers, each of which is formed by the same digit repeated the same number of times as in the table below.

Around the year 1300, a book was written with the title (in Arabic) *Condensing Arithmetic Operations*, in which the following calculations were provided:

Number of 1s	The Equal Factors		Product
1	$1 \cdot 1$	=	1
2	$11 \cdot 11$	=	121
3	$111 \cdot 111$	=	12,321
4	$1{,}111 \cdot 1{,}111$	=	1,234,321
. . .	. . .		. . .
9	$111{,}111{,}111 \cdot 111{,}111{,}111$	=	12,345,678,987,654,321

One would expect that this pattern of palindromic numbers would continue. However, if you look below, with the number generated by squaring the number with ten 1s, the pattern is destroyed. Thus, to generalize here, we would be once again making a mistake.

Number of 1s	The Equal Factors		Products
1	1 · 1	=	1
2	11 · 11	=	121
3	111 · 111	=	12,321
4	1,111 · 1,111	=	1,234,321
5	11,111 · 11,111	=	123,454,321
6	111,111 · 111,111	=	12,345,654,321
7	1,111,111 · 1,111,111	=	1,234,567,654,321
8	11,111,111 · 11,111,111	=	123,456,787,654,321
9	111,111,111 · 111,111,111		12,345,678,987,654,321
10	1,111,111,111 · 1,111,111,111	=	1,234,567,**900,98**7,654,321

WHEN CANCELLATIONS ARE MISTAKES, AND WHEN THEY ARE NOT!

This mistake has been sometimes referred to as a "howler." It is just the kind of mistake that can surely bring us to wonder! Consider the following, where we are asked to reduce the fraction $\frac{16}{64}$, and we simply cross out the 6s, which, strangely enough, gives us the right result: $\frac{16}{64}=\frac{1\not{6}}{\not{6}4}=\frac{1}{4}$. We can apply this method to the following as well:

To reduce the fraction $\frac{26}{65}$, we simply cross out the 6s to get the right answer: $\frac{26}{65}=\frac{2\not{6}}{\not{6}5}=\frac{2}{5}$.

To reduce the fraction $\frac{19}{95}$, we simply cross out the 9s to get the right answer: $\frac{19}{95}=\frac{1\not{9}}{\not{9}5}=\frac{1}{5}$.

To reduce the fraction $\frac{49}{98}$, we simply cross out the 9s to get the right answer: $\frac{49}{98}=\frac{4\not{9}}{\not{9}8}=\frac{4}{8}\left(=\frac{1}{2}\right)$.

Naturally, this can be done for all two-digit multiples of 11 ($\frac{11}{11}$, $\frac{22}{22}$, . . .), but that is as much as we can apply this silly method for two-digit numbers. One then wonders why this simple (or silly) method cannot always be used.

Sometimes an erroneous method can lead (just by coincidence) to a correct result, as in the cases above. The danger, of course, is that we must not generalize this method. These are the only two-digit examples of this mistaken procedure that lead to a correct reduction of a fraction.[2]

An arithmetic explanation for why this works can be seen from the following calculation:

$$\frac{16}{64}=\frac{1\cdot 10+6}{10\cdot 6+4}=\frac{6\cdot\frac{16}{6}}{6\cdot\frac{64}{6}}=\frac{6\cdot\frac{8}{3}}{6\cdot\frac{32}{3}}=\frac{8}{32}=\frac{1}{4}\text{; therefore, }\frac{16}{64}=\frac{1\cdot 10+\not{6}}{10\cdot\not{6}+4}=\frac{1}{4}.$$

For those readers with a good working knowledge of elementary algebra, we can "explain" this situation, and show that the four fractions above are the *only* ones (composed of two-digit numbers) where this type of cancellation will hold true. (Since this explanation uses only elementary algebra.)

We begin by considering the fraction $\frac{10x+a}{10a+y}$.

The above four cancellations were such that when canceling the a's the fraction was equal to $\frac{x}{y}$.

Therefore, $\frac{10x+a}{10a+y}=\frac{x}{y}$.

This yields: $y(10x+a)=x(10a+y)$,

$$10xy+ay=10ax+xy,$$

$$9xy+ay=10ax.$$

And so $y=\dfrac{10ax}{9x+a}$.

At this point we shall inspect this equation. It is necessary that x, y, and a are integers, since they were digits in the numerator and denominator of a fraction. It is now our task to find the values of a and x for which y will also be integral.

To avoid a lot of algebraic manipulation, you will want to set up a chart, which will generate values of y from $y=\frac{10ax}{9x+a}$. Remember that x, y, and a must be single-digit integers. Below is a portion of the table you will be constructing. Notice that the cases where $x = a$ are excluded since $\frac{x}{a}=1$.

x/a	1	2	3	4	5	6	7	8	9
1		$\frac{20}{11}$	$\frac{30}{12}$	$\frac{40}{13}$	$\frac{50}{14}$	$\frac{60}{15}=4$	$\frac{70}{16}$	$\frac{80}{17}$	$\frac{90}{18}=5$
2	$\frac{20}{19}$		$\frac{60}{21}$	$\frac{80}{22}$	$\frac{100}{23}$	$\frac{120}{24}=5$	$\frac{140}{25}$	$\frac{160}{26}$	$\frac{180}{27}$
3	$\frac{30}{28}$	$\frac{60}{29}$		$\frac{120}{31}$	$\frac{150}{32}$	$\frac{180}{33}$	$\frac{210}{34}$	$\frac{240}{35}$	$\frac{270}{36}$
4	$\frac{40}{37}$	$\frac{80}{38}$	$\frac{120}{39}$		$\frac{200}{41}$	$\frac{240}{42}$	$\frac{280}{43}$	$\frac{320}{44}$	$\frac{360}{45}=8$
5	$\frac{50}{46}$	$\frac{100}{47}$	$\frac{150}{48}$	$\frac{200}{49}$		$\frac{300}{51}$	$\frac{350}{52}$	$\frac{400}{53}$	$\frac{450}{54}$
6	$\frac{60}{55}$	$\frac{120}{56}$	$\frac{180}{57}$	$\frac{240}{58}$	$\frac{300}{59}$		$\frac{420}{61}$	$\frac{480}{62}$	$\frac{540}{63}$
7	$\frac{70}{64}$	$\frac{140}{65}$	$\frac{210}{66}$	$\frac{280}{67}$	$\frac{350}{68}$	$\frac{420}{69}$		$\frac{560}{71}$	$\frac{630}{72}$
8	$\frac{80}{73}$	$\frac{160}{74}$	$\frac{240}{75}$	$\frac{320}{76}$	$\frac{400}{77}$	$\frac{480}{78}$	$\frac{560}{79}$		$\frac{720}{81}$
9	$\frac{90}{82}$	$\frac{180}{83}$	$\frac{270}{84}$	$\frac{360}{85}$	$\frac{450}{86}$	$\frac{540}{87}$	$\frac{630}{88}$	$\frac{720}{89}$	

The portion of the chart, pictured above, already generated two of the four integral values of y; that is, when $x = 1$, and $a = 6$, then $y = 4$; and when $x = 2$ and $a = 6$, then $y = 5$. These values yield the fractions $\frac{16}{64}$ and $\frac{26}{65}$, respectively. The remaining two integral values of y will be obtained when $x = 1$ and $a = 9$, yielding $y = 5$; and when $x = 4$ and $a = 9$, yielding $y = 8$. These generate the fractions $\frac{19}{95}$ and $\frac{49}{98}$, respectively. This should convince you that there are only four such fractions composed of two-digit numbers.

You may now wonder if there are fractions composed of numerators and denominators of more than two digits where this strange type of cancellation holds true.

What follows are some examples of three digit numbers with this sort of strange cancellation can work.

$$\frac{199}{995}=\frac{1\cancel{99}}{\cancel{99}5}\left(=\frac{1}{5}\right),\ \frac{266}{665}=\frac{2\cancel{66}}{\cancel{66}5}\left(=\frac{2}{5}\right),\ \frac{124}{217}=\frac{\cancel{1}24}{2\cancel{1}7}\left(=\frac{4}{7}\right),\ \frac{103}{206}=\frac{1\cancel{0}3}{2\cancel{0}6}=\frac{13}{26}\left(=\frac{1}{2}\right),$$

$$\frac{495}{990}=\frac{4\cancel{9}5}{\cancel{9}90}=\frac{45}{90}\left(=\frac{1}{2}\right),\ \frac{165}{660}=\frac{1\cancel{6}5}{6\cancel{6}0}=\frac{15}{60}\left(=\frac{1}{4}\right),\ \frac{127}{762}=\frac{12\cancel{7}}{\cancel{7}62}\left(=\frac{1}{6}\right),\text{ and}$$

$$\frac{143185}{1701856}=\frac{143\cancel{18}5}{170\cancel{18}56}=\frac{1435}{17056}\left(=\frac{35}{416}\right).$$

Try this type of cancellation with $\frac{499}{998}$. You should find that $\frac{499}{998}=\frac{4}{8}=\frac{1}{2}$.

This can be extended as follows:

$$\frac{19999}{99995}=\frac{19999}{99995}=\frac{1999}{9995}=\frac{199}{995}=\frac{19}{95}=\frac{1}{5}, \text{ that is, } \frac{19999}{99995}=\frac{19999}{99995}=\frac{1}{5}, \text{ or}$$

$$\frac{26666}{66665}=\frac{26666}{66665}=\frac{2666}{6665}=\frac{266}{665}=\frac{26}{65}=\frac{2}{5}, \text{ that is, } \frac{26666}{66665}=\frac{26666}{66665}=\frac{2}{5}.$$

A pattern is now emerging, and you may realize that:

$$\frac{49}{98}=\frac{499}{998}=\frac{4999}{9998}=\frac{49999}{99998}=\ldots$$
$$\frac{16}{64}=\frac{166}{664}=\frac{1666}{6664}=\frac{16666}{66664}=\frac{166666}{666664}=\ldots$$
$$\frac{19}{95}=\frac{199}{995}=\frac{1999}{9995}=\frac{19999}{99995}=\frac{199999}{999995}=\ldots$$
$$\frac{26}{65}=\frac{266}{665}=\frac{2666}{6665}=\frac{26666}{66665}=\frac{266666}{666665}=\ldots$$

Enthusiastic readers may wish to justify these extensions of the original fractions of this sort. Readers who, at this point, have a further desire to seek out additional fractions that permit this strange cancellation should consider the following fractions. They should verify the legitimacy of this strange cancellation and then set out to discover more such fractions.

$$\frac{3\cancel{3}2}{8\cancel{3}0}=\frac{32}{80}=\frac{2}{5}$$

$$\frac{3\cancel{8}5}{8\cancel{8}0}=\frac{35}{80}=\frac{7}{16}$$

$$\frac{1\cancel{3}8}{\cancel{3}45}=\frac{18}{45}=\frac{2}{5}$$

$$\frac{2\cancel{7}5}{7\cancel{7}0}=\frac{25}{70}=\frac{5}{14}$$

$$\frac{1\cancel{6}\cancel{3}}{\cancel{3}2\cancel{6}}=\frac{1}{2}$$

However be careful, because $\frac{16\cancel{3}}{\cancel{3}26}\neq\frac{1}{2}$ and $\frac{1\cancel{6}3}{32\cancel{6}}\neq\frac{1}{2}$.

Aside from providing an algebraic application, which can be used to introduce a number of important topics in a motivational way, this topic can also provide some recreational activities. Here are some more of these weird fractions—clearly a mistaken procedure—yet leading to correct results!

$$\frac{4\cancel{8}\cancel{4}}{\cancel{8}\cancel{4}7}=\frac{4}{7} \qquad \frac{\cancel{5}\cancel{4}5}{6\cancel{5}\cancel{4}}=\frac{5}{6} \qquad \frac{\cancel{4}\cancel{2}4}{7\cancel{4}\cancel{2}}=\frac{4}{7} \qquad \frac{24\cancel{9}}{\cancel{9}96}=\frac{24}{96}=\frac{1}{4}$$

$$\frac{4\cancel{8}\cancel{4}\cancel{8}\cancel{4}}{\cancel{8}\cancel{4}\cancel{8}\cancel{4}7}=\frac{4}{7} \qquad \frac{\cancel{5}\cancel{4}\cancel{5}\cancel{4}5}{6\cancel{5}\cancel{4}\cancel{5}\cancel{4}}=\frac{5}{6} \qquad \frac{\cancel{4}\cancel{2}\cancel{4}\cancel{2}4}{7\cancel{4}\cancel{2}\cancel{4}\cancel{2}}=\frac{4}{7}$$

$$\frac{\cancel{3}\cancel{2}\cancel{4}3}{4\cancel{3}\cancel{2}\cancel{4}}=\frac{3}{4} \qquad \frac{\cancel{6}\cancel{4}\cancel{8}6}{8\cancel{6}\cancel{4}\cancel{8}}=\frac{6}{8}=\frac{3}{4}$$

$$\frac{14\cancel{7}\cancel{1}\cancel{4}}{\cancel{7}\cancel{1}\cancel{4}68}=\frac{14}{68}=\frac{7}{34} \qquad \frac{\cancel{8}\cancel{7}\cancel{8}\cancel{0}\cancel{4}8}{9\cancel{8}\cancel{7}\cancel{8}\cancel{0}\cancel{4}}=\frac{8}{9}$$

$$\frac{1\cancel{4}\cancel{2}\cancel{8}\cancel{5}\cancel{7}\cancel{1}}{\cancel{4}\cancel{2}\cancel{8}\cancel{5}\cancel{7}\cancel{1}3}=\frac{1}{3} \qquad \frac{2\cancel{8}\cancel{5}\cancel{7}\cancel{1}\cancel{4}\cancel{2}}{\cancel{8}\cancel{5}\cancel{7}\cancel{1}\cancel{4}\cancel{2}6}=\frac{2}{6}=\frac{1}{3} \qquad \frac{3\cancel{4}\cancel{6}\cancel{1}\cancel{5}\cancel{3}\cancel{8}}{\cancel{4}\cancel{6}\cancel{1}\cancel{5}\cancel{3}\cancel{8}4}=\frac{3}{4}$$

$$\frac{\cancel{7}\cancel{6}\cancel{7}\cancel{1}\cancel{2}\cancel{3}\cancel{2}\cancel{8}7}{8\cancel{7}\cancel{6}\cancel{7}\cancel{1}\cancel{2}\cancel{3}\cancel{2}\cancel{8}}=\frac{7}{8} \qquad \frac{\cancel{3}\cancel{2}\cancel{4}\cancel{3}\cancel{2}\cancel{4}\cancel{3}\cancel{2}\cancel{4}3}{4\cancel{3}\cancel{2}\cancel{4}\cancel{3}\cancel{2}\cancel{4}\cancel{3}\cancel{2}\cancel{4}}=\frac{3}{4}$$

$$\frac{\cancel{1}\cancel{0}\cancel{2}\cancel{5}\cancel{6}\cancel{4}1}{4\cancel{1}\cancel{0}\cancel{2}\cancel{5}\cancel{6}\cancel{4}}=\frac{1}{4} \qquad \frac{\cancel{3}\cancel{2}\cancel{4}\cancel{3}\cancel{2}\cancel{4}3}{4\cancel{3}\cancel{2}\cancel{4}\cancel{3}\cancel{2}\cancel{4}}=\frac{3}{4} \qquad \frac{4\cancel{5}\cancel{7}\cancel{1}\cancel{4}\cancel{2}\cancel{8}}{\cancel{5}\cancel{7}\cancel{1}\cancel{4}\cancel{2}\cancel{8}5}=\frac{4}{5}$$

$$\frac{4\cancel{8}\cancel{4}\cancel{8}\cancel{4}\cancel{8}\cancel{4}}{\cancel{8}\cancel{4}\cancel{8}\cancel{4}\cancel{8}\cancel{4}7}=\frac{4}{7} \qquad \frac{5\cancel{9}\cancel{5}\cancel{2}\cancel{3}\cancel{8}\cancel{0}}{\cancel{9}\cancel{5}\cancel{2}\cancel{3}\cancel{8}\cancel{0}8}=\frac{5}{8} \qquad \frac{\cancel{4}\cancel{2}\cancel{8}\cancel{5}\cancel{7}\cancel{1}4}{6\cancel{4}\cancel{2}\cancel{8}\cancel{5}\cancel{7}\cancel{1}}=\frac{4}{6}=\frac{2}{3}$$

$$\frac{\cancel{5}\cancel{4}\cancel{5}\cancel{4}\cancel{5}\cancel{4}5}{6\cancel{5}\cancel{4}\cancel{5}\cancel{4}\cancel{5}\cancel{4}}=\frac{5}{6} \qquad \frac{6\cancel{9}\cancel{2}\cancel{3}\cancel{0}\cancel{7}\cancel{6}}{\cancel{9}\cancel{2}\cancel{3}\cancel{0}\cancel{7}\cancel{6}8}=\frac{6}{8}=\frac{3}{4} \qquad \frac{\cancel{4}\cancel{2}\cancel{4}\cancel{2}\cancel{4}\cancel{2}4}{7\cancel{4}\cancel{2}\cancel{4}\cancel{2}\cancel{4}\cancel{2}}=\frac{4}{7}$$

$$\frac{\cancel{5}\cancel{3}\cancel{8}\cancel{4}\cancel{6}\cancel{1}5}{7\cancel{5}\cancel{3}\cancel{8}\cancel{4}\cancel{6}\cancel{1}}=\frac{5}{7} \qquad \frac{\cancel{2}\cancel{0}\cancel{5}\cancel{1}\cancel{2}\cancel{8}2}{8\cancel{2}\cancel{0}\cancel{5}\cancel{1}\cancel{2}\cancel{8}}=\frac{2}{8}=\frac{1}{4} \qquad \frac{\cancel{3}\cancel{1}\cancel{1}\cancel{6}\cancel{8}\cancel{8}3}{8\cancel{3}\cancel{1}\cancel{1}\cancel{6}\cancel{8}\cancel{8}}=\frac{3}{8}$$

$$\frac{\cancel{6}\cancel{4}\cancel{8}\cancel{6}\cancel{4}\cancel{8}6}{8\cancel{6}\cancel{4}\cancel{8}\cancel{6}\cancel{4}\cancel{8}}=\frac{6}{8}=\frac{3}{4} \qquad \frac{4\cancel{8}\cancel{4}\cancel{8}\cancel{4}\cancel{8}\cancel{4}\cancel{8}\cancel{4}}{\cancel{8}\cancel{4}\cancel{8}\cancel{4}\cancel{8}\cancel{4}\cancel{8}\cancel{4}7}=\frac{4}{7}$$

Mathematics continues to hold some hidden treasures wrapped up within "mistaken procedures."

A. P. Darmoryad has extended our appreciation for these strange calculations with the following example.[3] Yet we must be careful, because this is wrong!

$$\frac{4251935345}{91819355185}=\frac{425\cancel{1}\cancel{9}\cancel{3}\cancel{5}345}{918\cancel{1}\cancel{9}\cancel{3}\cancel{5}5185}=\frac{425345}{9185185}$$

However, in a very strange way, if we reduce this first fraction by dividing both numerator and the nominator by 5, it does work as follows:

$$\frac{4251935345}{91819355185}=\frac{850\left[387\right]069}{1836\left[387\right]1037}\approx\frac{850069}{18361037}\approx\frac{425345}{9185185}.$$

Reducing fractions simply by canceling can lead not only to errors but also to some creative results.

Here are a few examples of weird cancellation leading to correct answers:

$$\frac{19+2\cdot 1}{1+9+1\cdot 2}=\frac{\cancel{19}+2\cdot 1}{\cancel{1+9}+1\cdot 2}=\frac{21}{12},\quad \frac{28+3\cdot 1}{2+8+1\cdot 3}=\frac{\cancel{28}+3\cdot 1}{\cancel{2+8}+1\cdot 3}=\frac{31}{13},$$

$$\frac{37+4\cdot 1}{3+7+1\cdot 4}=\frac{\cancel{37}+4\cdot 1}{\cancel{3+7}+1\cdot 4}=\frac{41}{14},\quad \frac{46+5\cdot 1}{4+6+1\cdot 5}=\frac{\cancel{46}+5\cdot 1}{\cancel{4+6}+1\cdot 5}=\frac{51}{15},$$

$$\frac{55+6\cdot 1}{5+5+1\cdot 6}=\frac{\cancel{5}5+6\cdot 1}{\cancel{5+}5+1\cdot 6}=\frac{61}{16},\ \frac{64+7\cdot 1}{6+4+1\cdot 7}=\frac{\cancel{6}4+7\cdot 1}{\cancel{6+}4+1\cdot 7}=\frac{71}{17},$$

$$\frac{73+8\cdot 1}{7+3+1\cdot 8}=\frac{\cancel{7}3+8\cdot 1}{\cancel{7+}3+1\cdot 8}=\frac{81}{18},\ \frac{82+9\cdot 1}{8+2+1\cdot 9}=\frac{\cancel{8}2+9\cdot 1}{\cancel{8+}2+1\cdot 9}=\frac{91}{19}.$$

Care must also be taken for correct cancellations when we look at the values covering the general case. Consider the following equation:

$$\frac{(1+x)^2}{1-x^2}=\frac{1+x}{1-x}.$$

This is *almost* correct. Why do we say "almost"?

Clearly, if we take the left side and reduce the fraction appropriately, we will get the following:

$$\frac{(1+x)^2}{1-x^2}=\frac{(1+x)(1+x)}{(1+x)(1-x)}=\frac{1+x}{1-x}.$$

We notice that the left side $\frac{(1+x)^2}{1-x^2}$ is not defined for $x = 1$, nor for $x = -1$, since it will produce a 0 in the denominator. The right side $\frac{1+x}{1-x}$ is not defined only for $x = 1$.

MISTAKEN WITH PERCENTS

A common arithmetic mistake is one that can be encountered in a store that may have just increased its prices by, say, 10 percent and then, after noticing a drop in sales, decreases its prices by 10 percent with the notion that this will restore the prices to their original level. This is false! Perhaps the best way of demonstrating this mistake is to begin with an item costing \$100 and increase the price 10 percent to get a new price of \$110. Then, by reducing this new price (\$110) by 10 percent, or \$11, you get a reduced price of \$99. This is an often-overlooked mistake in calculation.

A similar mistake is made when one is offered a 20 percent discount on an already 10 percent discounted item in a store. The mistake is to believe that you are actually getting a 30 percent discount. Let us once again take the \$100 item and discount it 10 percent. That will give us a price of \$90. Discounting the \$90 item by 20 percent gives us \$72—not

\$70, as a 30 percent discount at the start would have given us. This is a common mistake and one that can easily mislead consumers. You will also notice that it will make no difference if we had taken the 10 percent discount before the 20 percent discount. The result would still be \$72.

An interesting and quite unusual procedure is provided for entertainment and to provide a fresh look into this problem situation. Here is a mechanical method for obtaining a single percentage discount (or increase) equivalent to two (or more) successive discounts (or increases).

(1) Change each of the percents involved into decimal form:
0.20 and 0.10.

(2) Subtract each of these decimals from 1.00:
0.80 and 0.90 (for an increase, add to 1.00).

(3) Multiply these differences:
$0.80 \cdot 0.90 = 0.72$.

(4) Subtract this number (i.e., 0.72) from 1.00:
$1.00 - 0.72 = 0.28$, which represents the combined *discount.*

(If the result of step 3 is greater than 1.00, subtract 1.00 from it to obtain the percent of *increase.*)

This procedure will also work with more than two discounts and increases—used in combination as well.

Another common error in working with percents can be seen with a recently reported construction project. A parking lot is to get some shrubbery, which will require making each parking space a bit smaller. The spaces will be reduced by 4 percent in length and by 5 percent in width. It is, therefore, calculated that the area of each space will be reduced by ($4 \cdot 5 =$) 20 percent. This is a mistake! Consider the following, where a represents the length of the current parking space, and b represents its width. We begin by representing the area of a current parking space.

$A_{\text{original}} = a_{\text{original}} \cdot b_{\text{original}}$. Then the new parking space would be:

$$A_{\text{new}} = a_{\text{new}} \cdot b_{\text{new}} = \left(a_{\text{original}} - \frac{4}{100} \cdot a_{\text{original}}\right) \cdot \left(b_{\text{original}} - \frac{5}{100} \cdot b_{\text{original}}\right) = \frac{96}{100} \cdot a_{\text{original}}$$

$$\cdot \frac{95}{100} \cdot b_{\text{original}} = \frac{114}{125} \cdot a_{\text{original}} \cdot b_{\text{original}} = 0.912\, a_{\text{original}} \cdot b_{\text{original}} \approx 0.91\, A_{\text{original}}.$$

This shows us that the mistake was quite substantial in that the spaces were actually reduced by about 9 percent, not 20 percent!

This sort of mistake can happen in even more dramatic fashion. Consider the homeowner who wants to double the volume of his swimming pool and decides to simply double each of the dimensions of the pool. This is a very costly mistake. If we let a, b, and c represent the three dimensions of the pool: width, length, and depth, we would determine the pool's volume with: $V_{\text{original}} = a_{\text{original}} \cdot b_{\text{original}} \cdot c_{\text{original}}$. Then the enlarged pool would be:

$$V_{\text{new}} = a_{\text{new}} \cdot b_{\text{new}} \cdot c_{\text{new}} = 2a_{\text{original}} \cdot 2b_{\text{original}} \cdot 2c_{\text{original}}$$
$$= 8 \cdot a_{\text{original}} \cdot b_{\text{original}} \cdot c_{\text{original}} = 8V_{\text{original}}.$$

To this homeowner's surprise, the volume was not doubled, but rather it has become eight times the original volume. One might call this a rather significant mistake!

IGNORING FRACTIONS CAN (MISTAKENLY) LEAD TO A CORRECT ANSWER

Can we say that "annoying" fractions can simply be ignored? Can you imagine that the product:

$$\left(a+\frac{b}{c}\right) \cdot \left(x-\frac{y}{z}\right) = a \cdot x.$$

That is, can we just ignore the two fractions $\frac{b}{c}$ and $\frac{y}{z}$ and still get the correct answer? Let's take a look at the following calculation, where we will do just that—ignore the fractions!

$$\left(7+\frac{3}{7}\right)\cdot\left(4-\frac{3}{13}\right)=7\cdot 4=28$$

How is this possible? Just look at the calculation.

First we (correctly) multiply the two binomial expressions, and then simplify:

$$\left(7+\frac{3}{7}\right)\cdot\left(4-\frac{3}{13}\right)=7\cdot 4-7\cdot\frac{3}{13}+\frac{3}{7}\cdot 4-\frac{3}{7}\cdot\frac{3}{13}=28-\frac{21}{13}+\frac{12}{7}-\frac{9}{91}=28.$$

Or you can also get the correct answer by converting each parenthetical expression to an improper fraction and then multiply:

$$\left(\frac{52}{7}\right)\cdot\left(\frac{49}{13}\right)=28.$$

Yes, we arrived at that same answer as we did when we ignored the fraction portions of the binomial expressions.

The same weird result can be obtained with the following example: $\left(7+\frac{1}{2}\right)\cdot\left(5-\frac{1}{3}\right)=7\cdot 5=35$, which we can easily verify just as we did above. In case you are still not convinced, here is yet another example of eliminating the "annoying" fractions—making the same mistake—and still getting the correct answer:

$$\left(31+\frac{1}{2}\right)\cdot\left(21-\frac{1}{3}\right)=31\cdot 21=651.$$

And to make this point further, consider still another such example:

$$\left(6+\frac{1}{4}\right)\cdot\left(5-\frac{1}{5}\right)=6\cdot 5=30.$$

Yes, the above calculations also (strangely) lead to the correct answer. Wow, what is actually going on here? Might this be a delightful simplification for future calculations? However, much to everyone's disappointment, this is not true for all cases of this form. Yes, it would be a mistake if this were to be extended to all such mixed numbers. Take, for example, the following:

$$\left(7+\frac{1}{5}\right)\cdot\left(4-\frac{2}{3}\right)=24,$$

which is not equal to $7\cdot 4=28$, which you might have anticipated using the above "rule."

At this point you must be curious to know when the following is true:

$$\left(a+\frac{b}{c}\right) \cdot \left(x-\frac{y}{z}\right) = a \cdot x.$$

Let's consider the following product, where we can inspect the situation:

$$\left(7+\frac{1}{2}\right) \cdot \left(5-\frac{1}{3}\right) = 7 \cdot 5 = 35.$$

Consider each of the two binomials separately: $7+\frac{1}{2} = 15 \div 2$, and $5-\frac{1}{3} = 14 \div 3$.

Notice that 2 is a divisor of 14, and 3 is a divisor of 15. Any nonprime integers differing by a small amount in comparison with the divisors can be used.

This rather magnificent mistake can also be extended to three numbers, as shown with the following two examples:

$$\left(2 \cdot \frac{2}{13}\right) \cdot \left(6+\frac{1}{4}\right) \cdot \left(5+\frac{1}{5}\right) = 2 \cdot 6 \cdot 5 = 60$$

$$\left(2 \cdot \frac{2}{13}\right) \cdot \left(4+\frac{1}{6}\right) \cdot \left(5+\frac{1}{5}\right) = 2 \cdot 4 \cdot 5 = 40$$

The ambitious reader may want to find other such special products.

STRANGE EXPONENTIAL RELATIONSHIPS THAT CAN LEAD TO A MISTAKEN PROCESS

Consider the following calculation: $2^5 \cdot 9^2 = 32 \cdot 81 = 2{,}592$. Notice how the digits in the two bases (2, 9) and the two exponents (5, 2) line up in order in the product 2,592. Is there a rule we can conclude from this example? Or would it be a mistake to draw a generalized rule from this one case? Much to our disappointment, this is the only time such a pattern exists. This misleading relationship was first discovered by Underwood Dudeney (1937–). It is quite easy to find a counterexample: $2^2 \cdot 2^2 = 16 \neq 2{,}222$. In 1934, the recreational mathematics proponent Charles W. Trigg[4] showed that the only possible solution for such a relationship is the above example. That is, only when $a = d = 2$, $b = 5$ and $c = 9$, will the equation $a^b \cdot c^d = \overline{abcd}$, where $\overline{abcd}$ is the decimal number with the digits in this order. Or, to define it more precisely, this expression $\overline{abcd} = 10^3 \cdot a + 10^2 \cdot b + 10^1 \cdot c + 10^0 \cdot d$, where $a, b, c, d \in \{0, 1, 2, 3, \ldots, 9\}$ and $a \neq 0$.

With a different process (this time subtraction) we have $8^2 - 2^2 = 82 - 22 = 60$, which turns out to be correct, since $64 - 4 = 60$. Also: $9^2 - 1^2 = 92 - 12 = 80$, which, again, is correct, since $81 - 1 = 80$.

But please do not make the mistake by extending this to other such situations.

To further protect the reader from making mistakes from generalizing some lovely discoverable patterns, look at the following "rule":

"If we seek the square of the sum of the digits of a number, just drop the plus signs and the exponent." An example of this is $(8 + 1)^2 = 81$. This can also hold for cubing a number, as with the following example: $(5 + 1 + 2)^3 = 512$ $(= 8^3)$. What follows is a list of other such examples, where dropping the plus signs yields the correct answer.

(Sum of the Digits)^n		**Power**		**Number**
4^1	=	4^1	=	4
$(8 + 1)^2$	=	9^2	=	81
$(5 + 1 + 2)^3$	=	8^3	=	512
$(1 + 9 + 6 + 8 + 3)^3$	=	27^3	=	19,683
$(2 + 4 + 0 + 1)^4$	=	7^4	=	2,401
$(1 + 6 + 7 + 9 + 6 + 1 + 6)^4$	=	36^4	=	1,679,616
$(5 + 2 + 5 + 2 + 1 + 8 + 7 + 5)^5$	=	35^5	=	52,521,875
$(2 + 0 + 5 + 9 + 6 + 2 + 9 + 7 + 6)^5$	=	46^5	=	205,962,976
$(3 + 4 + 0 + 1 + 2 + 2 + 2 + 4)^6$	=	18^6	=	34,012,224
$(2 + 4 + 7 + 9 + 4 + 9 + 1 + 1 + 2 + 9 + 6)^6$	=	54^6	=	24,794,911,296
$(6 + 1 + 2 + 2 + 2 + 0 + 0 + 3 + 2)^7$	=	18^7	=	612,220,032

Unfortunately, once again, generalizing this nifty procedure would be a huge mistake, as we can see from the following example: $(8 + 2)^2 = 10^2$

$= 100$ is not obtainable by dropping the plus sign and the exponent, as before, since $(8 + 2)^2 \neq 82$. You may wish to search for other examples where this weird scheme does work.

We can find other such nifty calculation "shortcuts" that can only mistakenly be generalized, but are nice to observe. In one case, we have a situation where there is a sum of digits that are all to the same power. Here we just drop the exponents and the plus signs to get the correct answer: $1^3 + 5^3 + 3^3 = 153$ $(= 1 + 125 + 27)$.

This can be extended with sums of digits, each taken to various powers and taking us directly to the "correct" result.

(Sum of the Digits)^n		Number
$1^3 + 5^3 + 3^3$	=	153
$1^1 + 7^2 + 5^3$	=	175
$1^1 + 3^2 + 0^3 + 6^4$	=	1,306
$8^4 + 2^4 + 0^4 + 8^4$	=	8,208
$4^5 + 1^5 + 5^5 + 1^5$	=	4,151
$3^3 + 4^4 + 3^3 + 5^5$	=	3,435
$2^1 + 6^2 + 4^3 + 6^4 + 7^5 + 9^6 + 8^7$	=	2,646,798

Above all, bear in mind that to extend this false rule would be an embarrassing mistake, as we can see from the following: $1^1 + 2^2 + 3^3 = 1 + 4 + 27 = 32 \neq 123$.

We must always be careful when we discover some amazing relationships, as they may be just that and not necessarily generalizable. Remember, by definition, an exception can never be generalized, or else a colossal mistake will result.

The same kind of juggling of numbers and exponents to produce unusual results can be admired in the following examples:

$1 + 5 + 8 + 12 = 26 = 2 + 3 + 10 + 11$, and now look at the sum of the squares of these digits:

$1^2 + 5^2 + 8^2 + 12^2 = 234 = 2^2 + 3^2 + 10^2 + 11^2$.

Amazingly, this can be extended for these digits taken to the power of 3, as follows:

$1^3 + 5^3 + 8^3 + 12^3 = 2{,}366 = 2^3 + 3^3 + 10^3 + 11^3$.

Although we want to be cautious, and not to mistakenly set up any form of generalization, we offer the following for entertainment:

$1 + 6 + 7 + 8 + 14 + 15 = 51 = 2 + 3 + 9 + 10 + 11 + 16$

$1^2 + 6^2 + 7^2 + 8^2 + 14^2 + 15^2 = 571 = 2^2 + 3^2 + 9^2 + 10^2 + 11^2 + 16^2$

$1^3 + 6^3 + 7^3 + 8^3 + 14^3 + 15^3 = 7{,}191 = 2^3 + 3^3 + 9^3 + 10^3 + 11^3 + 16^3$

$1^4 + 6^4 + 7^4 + 8^4 + 14^4 + 15^4 = 96{,}835 = 2^4 + 3^4 + 9^4 + 10^4 + 11^4 + 16^4$.

DECIMAL FRACTIONS WITH PITFALLS

Let us recall that the designation: $0.\overline{6} = 0.666\ldots$ represents an infinite repeating decimal. Can we then conclude that $0.\overline{6} \cdot 0.\overline{3} = 0.\overline{18}$? To see if this is a mistaken process, we will change each of these infinite repeating decimals to fraction form:

$$0.\overline{6} = \frac{2}{3}, \text{ and } 0.\overline{3} = \frac{1}{3}.$$

Their product is then: $0.\overline{6} \cdot 0.\overline{3} = \frac{2}{3} \cdot \frac{1}{3} = \frac{2}{9} = 0.222\ldots = 0.\overline{2}$, and *not* $0.\overline{18}$. Does this imply that we should be rounding this to $0.\overline{2}$? To answer this question, we will evaluate $0.\overline{18}$ by changing it to fraction form.

We begin by letting $x = 0.\overline{18}$.

Therefore, $100x = 18.\overline{18}$, then, subtracting the two equations, we get $100x - x = 18.\overline{18} - 0.\overline{18} = 18$. Thus, $99x = 18$, and then $x = \frac{18}{99} = \frac{2}{11} = 0.\overline{18}$, which is *not* equal to $0.\overline{2} = \frac{2}{9}$. This is a mistake to be avoided!

WATCH FOR MISTAKEN IDENTITIES

It is pretty common knowledge that the $\sqrt{2}$ is irrational. When asked why that is the case, a common answer is that the decimal expansion shows no pattern of repeating. That would be a correct answer. However, that answer could also lead to a mistaken characterization of a number that may not be irrational. On a normal calculator, $\sqrt{2}$ would show as 1.4142136. Would this be enough information to determine whether or not there is a pattern? Not really. With the aid of a computer, we can take $\sqrt{2}$ to 100-place accuracy to get: 1.41421356237309504880168872420969807856967187537694807317667973799073247846210703885038753432764157 2, which does not show any pattern of repetition. Would this, then, be conclusive evidence for its irrationality? Before we answer that question, let's consider the following.

When we take the decimal expansion of the fraction $\frac{1}{7}$, we get: $\frac{1}{7}$ = 0.142857 **142857** 142857 **142857** . . . = $0.\overline{142857}$. By definition, a fraction is rational, and this is also seen by a repeating pattern in the decimal equivalent. However, this is where a mistake can be made. Consider the 100-place decimal expansion of the fraction $\frac{1}{109}$:

$\frac{1}{109}$ = 0.0091743119266055045871559633027522935779816513761467 8
899082568807339449541284403669724770642201834862 3

Using our test of a pattern to be recognized in the decimal expansion of a fraction (which we know is rational—by definition), we could easily make a mistake, since there is no pattern to be seen at this point. However, if we expand this fraction to 110 decimal places, we would get the following, where a repetition appears to be emerging with the last two digits, 91:

$\frac{1}{109}$ = 0.00**91**7431192660550458715596330275229357798165137614 67
88990825688073394495412844036697247706422018348623853 2
11 00**91**

Once again, with the assistance of a computer, we will expand the fraction 220 places:

$\frac{1}{109}$ = 0.0091743119266055045871559633027522935779816513761467
88990825688073394495412844036697247706422018348623853 2
1100**9174311926605504587155963302752293577981651376146 7**
8899082568807339449541284403669724770642201834862385 32
11009174.

We notice that there is a period of 108 decimal places before repetition occurs. (Advanced readers will recognize that the period of repetition can be at most $109 - 1 = 108$ decimal places.)

What does this, in effect, tell us? Essentially, we cannot use the argument of repeating decimal places as a criterion for determining irrationality because we are limited in our ability to expand the number in question. There are simple algebraic methods for determining the irrationality of a number.

DIMENSIONAL MISTAKES

On September 23, 1999, communication with the Mars Climate Orbiter was lost as the spacecraft went into orbital insertion, because the ground-based computer software produced output in English units of pound/seconds instead of the specified metric units of Newton/seconds. The spacecraft encountered Mars at an improperly low altitude, causing it to incorrectly enter the upper atmosphere and disintegrate. Dimensional errors can cause some curious mistakes in mathematics.

As we look into this sort of mistake, it will be helpful to consider a similar problem that uses different units. To keep it as similar as possible to our current problem, let's use a familiar system that also has a unit of measure that breaks down naturally into 100 subunits; in this case, the metric system fits the bill. We know that 1 meter is composed of 100 centimeters; in mathematical notation 1 m = 100 cm. Following the logic of the above equations, we continue: $1 \text{ m} = 100 \text{ cm} = (10 \text{ cm})^2 = \ldots.$

But wait! That isn't the case at all! In fact, $(10 \text{ cm})^2 = 100$ square cm, not 100 cm!

We must always keep in mind the type of units with which we are working: linear units, square units, or cubic units, and so on.

A mistake here could lead to the following conundrums:

$1¢ = \$0.01 = (\$0.10)^2 = (10¢)^2 = 100¢ = \$1$. Does that indicate that one cent equals one dollar?

Or, doing this in the reverse order, we get the same result:

$\$1 = 100¢ = (10¢)^2 = (\$0.10)^2 = \$0.01 = 1¢.$

Where is the money disappearing to? Have we really shown that \$1 = 1¢? Where is the mistake?

We can do the same thing with a quarter:

$25¢ = \$0.25 = (\$0.5)^2 = (50¢)^2 = 2{,}500¢ = \$25.$

And we can show in a similar way that \$1 = \$100 as follows:

$\$1 = (\$1)^2 = (100¢)^2 = 10{,}000¢ = \$100.$

We can even do the same thing with square roots, as shown here:

$5¢ = \sqrt{25¢} = \sqrt{\frac{1}{4}\$} = \sqrt{\frac{1}{2}\$ \cdot \frac{1}{2}\$} = \sqrt{50¢ \cdot 50¢} = \sqrt{2500¢} = 50¢.$

We can also show that \$1 = 10¢ in the following way:

We begin by dividing both sides by 100 to get: $\frac{\$1}{100} = \frac{100¢}{100}$, or $\frac{\$1}{100} = 1¢.$

When we take the square root of both sides, we get: $\sqrt{\frac{\$1}{100}} = \sqrt{1¢}$, or $\frac{\$1}{10} = 1¢.$

Multiplying both sides by 10 gives us: \$1 = 10¢.

So, then, where are the errors? We can't simply allow the dollar sign to escape the square root unscathed. It, too, must have a square root taken. But $\sqrt{\$}$ is not a unit that we know how to handle, nor is $\sqrt{¢}$—both are nonsensical as units of measure!

Changing the formulation of the problem reveals that the mistake was attributable to our initial mistreatment of the units of measure; since there is no such thing as "square cents" or "square dollars" as there are "square feet" and "square centimeters," our intuition passed right by that mistake. Putting the parentheses in the correct place, we see that

$$\$1 = 100¢ = (10)^2¢ = (10)^2\ ¢ \cdot \frac{\$1}{100¢} = \frac{\$(10)^2}{100} = \$1.$$

The main difference here is in the fourth step, where we use the unit conversion between dollars and cents, canceling the cents sign, and thereby return to the expected value of \$1.

In subjects like physics, on the other hand, we regularly encounter units of measure such as the meter per second-squared ($\frac{m}{s^2}$, not to be confused with $\left(\frac{m}{s}\right)^2$), which is the international standard unit of acceleration.

In a hypothetical problem involving the multiplication of acceleration by a distance, we would see acceleration times distance, which is expressed as:

$$\frac{m}{s^2} \cdot m = \frac{m^2}{s^2} = \left(\frac{m}{s}\right)^2,$$

and, taking the square root of this, we get $\sqrt{\left(\frac{m}{s}\right)^2} = \frac{m}{s}$.

This yields the meter per second, which is a unit of velocity. And indeed, it turns out that this is a valid physics formula—namely, the change in velocity of an object is equal to the square root of the product of the acceleration multiplied by the distance over which the acceleration is applied. So it turns out that paying attention to units is not only useful for avoiding mistakes; it can also be used as a technique for remembering practical principles!

TRY TO FIND THE MISTAKE HERE: A PARADOX

A customer walks into a bookshop and buys a book for $10. The next day, he returns to the bookshop and returns the book he bought the previous day. He then selects a book costing $20 and simply walks out with it. His reasoning is that he paid for the $10 book on the first day, and then returned the $10 book, thereby leaving $10.00 cash plus the $10.00 book behind. With this $20.00 credit, he then took a $20.00 book and considered it an even trade. Is this correct? If not, where is the error? There is obviously a subtle mistake made for the reader to discover. (Hint: Try doing this by replacing “a $10 book” with “two $5 bills.” The mistake should then become clear.)

THE PARADOX OF THE MISSING DOLLAR

Three men plan to spend one night in a hotel room. They pay $60.00 for the hotel room. Just as they were about to leave their room, the receptionist noticed that the cost for the room was $55.00 per night. The receptionist sends the bellhop to the room to return the $5.00 of overpayment. However, the bellhop decides to give each the three guests $1.00, and keeps the remaining $2.00 for himself. Therefore, to each other, each of the three guests has paid only $19.00 for the room. The sum of these three payments is therefore $3 \cdot \$19 = \$57$. This plus the $2.00 that the bellhop kept only totals to $59.00. Where is the missing dollar? Is there some mistake?

After a somewhat-bewildered reaction to this transaction, we offer the following explanation of the mistake: It is totally meaningless to add the $2.00 that the bellhop took to the $57.00 paid by the three men. The correct calculation is as follows: three men paid $57.00 for the room, of which $55.00 went to the receptionist and $2.00 went to the bellhop.

Another way of looking at this (unmistakenly), is to note that the three guests got a refund of $3.00, which when added to the $55.00 they originally paid for the room and the $2.00 the bellhop took, for a total of $60.00. Such calculating mistakes are not uncommon yet should not be accepted casually.

A MISTAKE IN AVERAGING RATES

A common mathematics mistake is one where we seek the average speed over a round-trip. For example, let's assume that the speed going to a destination is calculated at 60 mph and the return trip over the same route is calculated at 30 mph. The typical mistake here is to take the two speeds and treat them with equal weight. That is, to add them and then divide by 2 to get 45 mph. Although this would be the correct procedure for taking the arithmetic mean (the common average) of two quantities of *equal* value, it is not the correct way to treat two quantities of *unequal* value—in this case the rates taken over two *different* amounts of time.

This is the case for the two speeds being considered, since the speed of 30 mph is done for twice the time that the speed at 60 mph is done over the same distance. Consequently, the 30 mph speed must be given double the weight. Therefore, the correct average speed for the entire round-trip would be calculated as follows: $\frac{60+30+30}{3} = 40$ mph.

To get the average of speeds that are not as conveniently related as the two we used in the above example, we would employ the concept of a *harmonic mean*. The harmonic mean is defined as "the reciprocal of the average of the reciprocals" of the values being considered. That is, for two values a and b, the harmonic mean is

$$\frac{1}{\frac{\frac{1}{a}+\frac{1}{b}}{2}}=\frac{2}{\frac{1}{a}+\frac{1}{b}}=\frac{2ab}{a+b}.$$

For the above example, we can use the formula for the harmonic mean to get the average speed as follows:

$$\frac{2\cdot 60\cdot 30}{60+30}=40.$$

The harmonic mean becomes particularly useful when the numbers are not as convenient as those we used above. Suppose we now use as our round trip speeds 58 mph and 32 mph. Applying the harmonic mean formula to these two speeds, we get:

$$\frac{2\cdot 58\cdot 32}{58+32}=41.2\overline{4}.$$

Using the harmonic mean formula allows us also to find the average speed of more than two speeds. When we apply the definition of the harmonic mean, namely, the reciprocal of the average of the reciprocals, we get for three items the following:

$$\frac{1}{\frac{\frac{1}{a}+\frac{1}{b}+\frac{1}{c}}{3}}=\frac{3}{\frac{1}{a}+\frac{1}{b}+\frac{1}{c}}=\frac{3abc}{bc+ac+ab}.$$

And for four items:

$$\frac{1}{\frac{\frac{1}{a}+\frac{1}{b}+\frac{1}{c}+\frac{1}{d}}{4}}=\frac{4}{\frac{1}{a}+\frac{1}{b}+\frac{1}{c}+\frac{1}{d}}=\frac{4abcd}{bcd+acd+abd+abc}.$$

This can, of course, be expanded to five or more speeds (over the equal distances) that need to be averaged. With this relationship we can avoid one of the most common, albeit not magnificent, mistakes in mathematical calculations.

It should also be noted that a common procedure to find the average speed can also be done by finding the total distance and dividing it by the total time for the entire trip.

Mistakes in arithmetic can not only be annoying but also ruin some fine mathematics. Therefore, it is important to bear in mind the many arithmetic mistakes we have highlighted in this chapter.

CHAPTER 3

ALGEBRAIC MISTAKES

The adage that "one learns from mistakes" is particularly true when it comes to algebra. Naturally, one can make a calculation mistake, or perhaps a careless omission, or even overlook some of the basic rules of algebra. It is the last that is most important in our understanding of algebra. There are reasons why we have these rules. Most often a violation of these algebra rules will lead to absurd results. Even though they are oftentimes entertaining, these mistakes can also be very instructive. In this chapter, we will investigate all kinds of errors. It must be said that most errors give us a better understanding of mathematics, especially when we learn from the mistakes. But bear in mind, we will not be presenting silly mistakes like the following, where a student sees the correct value of $\lim_{x\to 0}\frac{8}{x} = \infty$, and then, when asked to find the value of $\lim_{x\to 0}\frac{5}{x}$, extends the given model to end up with $\lim_{x\to 0}\frac{5}{x} = \rotatebox{90}{5}$.

So let's now embark on our pursuit of some important algebraic mistakes.

As we mentioned earlier, possibly one of the most important rules in mathematics is that one is not allowed to divide by zero. Some even refer to this as the "eleventh commandment." There are times when division by zero is so well camouflaged that one violates this commandment without knowing it. It is interesting to see what happens when it is violated. Hopefully we will learn from each of these transgressions. What is interesting (or entertaining), is to discover when this rule—dividing by zero—has been violated, thus allowing us to arrive at ridiculous results. Let's consider a few of these mistakes.

DOES 1 = 2? A MISTAKE BASED ON DIVISION BY ZERO

If we square both sides of the equation $a = a$, we get $a^2 = a^2$. Then, subtracting a^2 from both sides of the equation, we have $a^2 - a^2 = a^2 - a^2$. We will factor the common term a on the left side and factor the difference of two squares on the right side to get $a\,(a - a) = (a + a)(a - a)$. As $a + a = 2a$, this can be rewritten as $a\,(a - a) = 2a\,(a - a)$. When we now divide both sides of this equation by $a\,(a - a)$, we get $1 = 2$. Where did we make the mistake? We have $a - a = 0$. Therefore, we have violated the important rule of not dividing by zero, resulting in an absurd statement, namely, $1 = 2$.

Here is another simple example of this sort of mistake—that of dividing by zero.

This time we will begin with the statement that $a = b$.

We then multiply both sides by b to get $a \cdot b = b \cdot b$, or $ab = b^2$.

Then subtract a^2 from both sides of the equation, so that $ab - a^2 = b^2 - a^2$.

Factoring the common factor on the left and the difference of two squares on the right, $a\,(b - a) = (b + a)(b - a)$.

Dividing both sides by $(b - a)$ gives us $a = b + a$.

However, since $a = b$ (which was given), $b = b + b$, or $b = 2b$, which, when we divide both sides by b, has us ending up with $1 = 2$.

In the following example, division by zero is more camouflaged, thereby making it a bit more difficult to detect.

TO SHOW THAT IF $a > b$, THEN $a = b$, THE MISTAKE OF DIVISION BY ZERO

We begin with $a > b$, which can be restated as $a = b + c$, where a, b, and c are positive numbers. We shall now multiply both sides of this equation by $a - b$ to get: $a^2 - ab = ab + ac - b^2 - bc$. We then subtract ac from both sides of the equation, giving us: $a^2 - ab - ac = ab - b^2 - bc$. Then, factoring both sides of the equations, we get $a(a - b - c) = b(a - b - c)$. Dividing both sides by $(a - b - c)$ leaves us with $a = b$.

How can it be that $a = b$, when we were told at the beginning that

$a > b$? Here we see that, once again, $(a - b - c)$ is equal to 0 because we began with the information that $a = b + c$. We have thus violated the rule that forbids dividing by zero.

TO SHOW THAT ALL INTEGERS ARE EQUAL BASED ON A MISTAKE: DIVISION BY ZERO

Once again we will use the mistake of dividing by zero—in a somewhat hidden fashion—to show this silly result. We begin by accepting the correct quotients that follow:

$$\frac{x-1}{x-1}=1.$$

Because $(x + 1)(x - 1) = x^2 - 1$, we get:

$$\frac{x^2-1}{x-1}=x+1.$$

Because $(x^2 + x + 1)(x - 1) = x^3 - 1$, we get:

$$\frac{x^3-1}{x-1}=x^2+x+1.$$

Because $(x^3 + x^2 + x + 1)(x - 1) = x^4 - 1$, we get:

$$\frac{x^4-1}{x-1}=x^3+x^2+x+1.$$

Because $(x^{n-1} + x^{n-2} + \ldots + x^3 + x^2 + x + 1)(x - 1) = x^n - 1$, we get:

$$\frac{x^n-1}{x-1}=x^n{-}1 + x^{n-2} + \ldots + x^3 + x^2 + x + 1.$$

Now suppose we let $x = 1$. The absolute values of the right sides of the above equations equal 1, 2, 3, 4, . . . , n. The left side of the above equations will all be the same, as they are in the form $\frac{1^n - 1}{1-1}$, and, therefore, all the right-side numbers must be equal, or $1 = 2 = 3 = 4 = \ldots = n$. Surely by now you will have realized that the denominators are all $1 - 1 = 0$. This cannot be permitted to exist, since if it did, then absurd conclusions, such as that all the numbers 1, 2, 3, . . ., n are equal. We can see here that $\frac{0}{0}$ cannot be a number, or all of these weird results would follow.

THE HIDDEN NEMESIS OF DIVIDING BY ZERO AND NOT NOTICING IT IMMEDIATELY

There are times when the division by zero is well camouflaged. Take for example the equation

$$\frac{3x-30}{11-x} = \frac{x+2}{x-7} - 4,$$

which allows the right side to be combined as $\frac{3x-30}{11-x} = \frac{x+2-4(x-7)}{x-7}$.

This can be then simplified to be $\frac{3x-30}{11-x} = \frac{3x-30}{7-x}$.

Since the numerators are equal, the denominators must also be equal, and, therefore, $11 - x = 7 - x$, or $11 = 7$. Quite an absurdity!

It doesn't appear that we divided by zero this time, and yet we ended up with an absurd result.

Had we solved the equation $\frac{3x-30}{11-x} = \frac{3x-30}{7-x}$ in the traditional way, we would find that $x = 10$, which would make the two numerators equal to zero. Still, that doesn't show that we divided by zero.

So we should consider the following: If $\frac{a}{b} = \frac{a}{c}$, and then we multiply both sides by bc to get $ac = ab$. Dividing both sides by a gives us $b = c$, which we expected. However, if $a = 0$, then this would not be valid, since we would have divided by zero.

Let us now return to the equation $\frac{3x-30}{11-x} = \frac{3x-30}{7-x}$, which led us to this absurd result. We found that $x = 10$. With that value of x, the numerators $3x - 30 = 0$, and, therefore, in this case we cannot equate the denominators. Notice how slyly division by zero hid from us to deliver a ridiculous result.

MORE OF THE HIDDEN NEMESIS: THE DIVISION BY ZERO

In a similar vein—but equally well hidden—we can show that $+1 = -1$. We begin with the equation

$$\frac{x+1}{p+q+1} = \frac{x-1}{p+q-1}.$$

By subtracting 1 from each side of this equation, we get:

$\frac{x+1}{p+q+1} - \frac{p+q+1}{p+q+1} = \frac{x-1}{p+q-1} - \frac{p+q-1}{p+q-1}$, which can be simplified to get:

$$\frac{x+1-(p+q+1)}{p+q+1} = \frac{x-1-(p+q-1)}{p+q-1}, \text{ or } \frac{x-p-q}{p+q+1} = \frac{x-p-q}{p+q-1}.$$

Since the numerators are equal, the denominators must also be equal, so that $p + q + 1 = p + q - 1$, or $+1 = -1$, an absurdity! Why did this happen? Might the previous example give a clue?

If you solve the original equation

$\frac{x+1}{p+q+1} = \frac{x-1}{p+q-1}$ for x, we find that $x = p + q$.

Therefore, we will have had the same situation as above, where the numerators of the two equal fractions $(x - p - q)$ were zero.

The initial equation

$$\frac{x+1}{p+q+1} = \frac{x-1}{p+q-1}$$

is not as general as we would at first imagine. It is relevant only for the case where $x = p + q$ and $p + q \neq \pm 1$.

To better understand this result we can look at a simpler version: From $\frac{a}{b} = \frac{a}{b}$, we cannot simply conclude that $\frac{a+c}{b+c} = \frac{a-c}{b-c}$, since that is only true if:

(1) $a = b$ and $(b + c)(b - c) \neq 0$, or

(2) $c = 0$ and $b \neq 0$.

In other words, we have to make sure that the denominator is not zero.

FINDING THE DIVISION BY ZERO BEFORE IT MISLEADS US

There are many examples of division-by-zero mistakes that follow a similar pattern. However, division by zero is usually camouflaged and sometimes difficult to find. There are terms that hid the zero so well that it can be easily overlooked—especially when you have no reason to suspect it being there. Let's consider the following example.

Suppose a term T_1 is divided by another term $T_2 = \sqrt{4-2\sqrt{3}} - \sqrt{3} + 1$; we would not be at all suspicious. However, as you will see in a moment, the term T_2 is of a nature that will violate our now-familiar "eleventh commandment," if we use it as a divisor. In fact, the term T_2 is equal to zero! Follow along the algebra and you will see that it equals zero.

Because $\sqrt{4-2\sqrt{3}} = \sqrt{3-2\sqrt{3}+1} = \sqrt{(\sqrt{3})^2-2\cdot1\cdot\sqrt{3}+1^2} = \sqrt{(\sqrt{3}-1)^2} = \sqrt{3}-1$, it then follows that $T_2 = \sqrt{4-2\sqrt{3}} - \sqrt{3} + 1 = 0$.

The zero divisor can be hidden even more, as shown in the following: $T_3 = \sqrt[3]{\sqrt{5}+2} + \sqrt[3]{\sqrt{5}-2} - \sqrt{5}$.

One may now wonder how we can show that $T_3 = 0$. Here is a hint that should help you show that $T_3 = 0$. Notice that $\sqrt[3]{\sqrt{5}+2} = \frac{\sqrt{5}+1}{2}$ and $\sqrt[3]{\sqrt{5}-2} = \frac{\sqrt{5}-1}{2}$, and $\frac{\sqrt{5}+1}{2} + \frac{\sqrt{5}-1}{2} = \sqrt{5}$.

AN ABSURD RESULT STEMMING FROM THE WELL-KNOWN MISTAKE BASED ON DIVISION BY ZERO

Before we begin this example, let's look at a basic principle from algebra. Consider the proportion $\frac{a}{b}=\frac{c}{d}$, only if $b \neq d$ (and $d \neq 0$). From this we can conclude that $\frac{a-c}{b-d}=\frac{c}{d}$. To show that this rule actually is true, we begin by recognizing that from the original proportion, $ad = bc$ (by cross multiplication). The cross multiplication for the second proportion (above) gives us the following: $(a-c)\cdot d = (b-d)\cdot c$, or $ad - cd = bc - dc$, which, when adding cd to both sides, gives us $ad = bc$. This is the same as what we had gotten for the first proportion. Now having established the rule above, we shall apply it to the following situation:

Given x, y, z and the proportion $\frac{3y-4z}{3y-8z}=\frac{3x-z}{3x-5z}$, we shall now apply the rule we established above to get:

$\frac{3y-4z-(3x-z)}{3y-8z-(3x-5z)}=\frac{3x-z}{3x-5z}$, and then $\frac{3y-4z-3x+z}{3y-8z-3x+5z}=\frac{3x-z}{3x-5z}$, which simplifies to:

$$\frac{3y-3z-3x}{3y-3z-3x}=1=\frac{3x-z}{3x-5z}.$$

It then follows that $3x - 5z = 3x - z$, which gives us $5 = 1$.

There must be something wrong, since we ended up with an absurd result.

The mistake here is a bit more difficult to find. The equation

$\frac{3y-4z}{3y-8z}=\frac{3x-z}{3x-5z}$ is satisfied when $x - y + z = 0$.

Now substituting $x = y - z$, we get:

$$\frac{3x-z}{3x-5z}=\frac{3(y-z)-z}{3(y-z)-5z}=\frac{3y-4z}{3y-8z}.$$

Using a similar substitution, we have $3y - 3z - 3x = 3y - 3z - 3(y - z) = 3y - 3z - 3y + 3z = 0$, and $3y - 3z - 3x = 3y - 8z + 5z - 3x = 3y - 8z - (3x - 5z) = 0$. Both are equal to zero. Thus the denominator of the fraction

$$\frac{3y-4z-(3x-z)}{3y-8z-(3x-5z)}$$

is equal to zero. This is an example of where division by zero is well camouflaged, and it exemplifies the kind of subtle mistakes that mathematicians have to be cautious about throughout their investigations.

ABSURD RESULTS BASED ON A MISTAKE OF INTERPRETATION

We are asked to solve the system of equations:

$$a + b = 1 \quad (1)$$
$$a + b = 2 \quad (2)$$

Our initial reaction is to notice that since the left sides of these equations are equal, then so must the right sides be equal. Thus, we find that $1 = 2$. "Proved!" Or is it?

We could also have embarked on this system of equations by subtracting the two equations—knowing that the difference of equals is also equal. When we subtract the first equation from the second equation, on the left side we would get 0 and on the right side we would get 1. Thus, 0 = 1. Once again, an absurd result.

To take this one step further, with this set of equations we could also show that 1 = –1. When we subtracted the first equation from the second equation, we got 0 = 1. Now, if we subtract the second equation from the first equation, we get 0 = –1. We can take this absurdity even one foolish step further.

Since 0 = 1 and 0 = –1, one could then conclude the 1 = –1, as both are equal to 0.

Our series of absurd conclusions above result from the fact that these two equations have no common solution. Were we to graph them, they would appear as two parallel lines—thus having no intersection or point in common. The mistake here was to embark on the two equations, seeking a solution and not recognizing immediately that there cannot be a solution when the two lines representing these equations are parallel, and, therefore, have no common point of intersection.

Some mistakes in algebra can be seen better graphically, as we will see in the following illustration. Consider the two equations: $5x + y = 15$, and $x = 4 - \frac{y}{5}$. If we substitute the value of x from the second equation into the first equation, we will get $5\left(4-\frac{y}{5}\right) + y = 15$. This simplifies to $20 - y + y = 15$, or $20 = 15$. Now, there must be something clearly wrong here. Where was the mistake? If we multiply both sides of the second of the two given equations by 5, we get $5x + y = 20$. Were we to graph these two equations, we would find them to be parallel and therefore to have no point of intersection; or, to put it another way, they have no common solution (see figure 3.1). Therefore, it makes no sense to try to solve these two equations simultaneously as we did above—thus clearly leading to an absurd result!

This time the parallelism of the two equations was not as obvious as in the first case above. Yet, to avoid such absurd results, we have to be cautious not to make some of the mistakes of interpretation shown here.

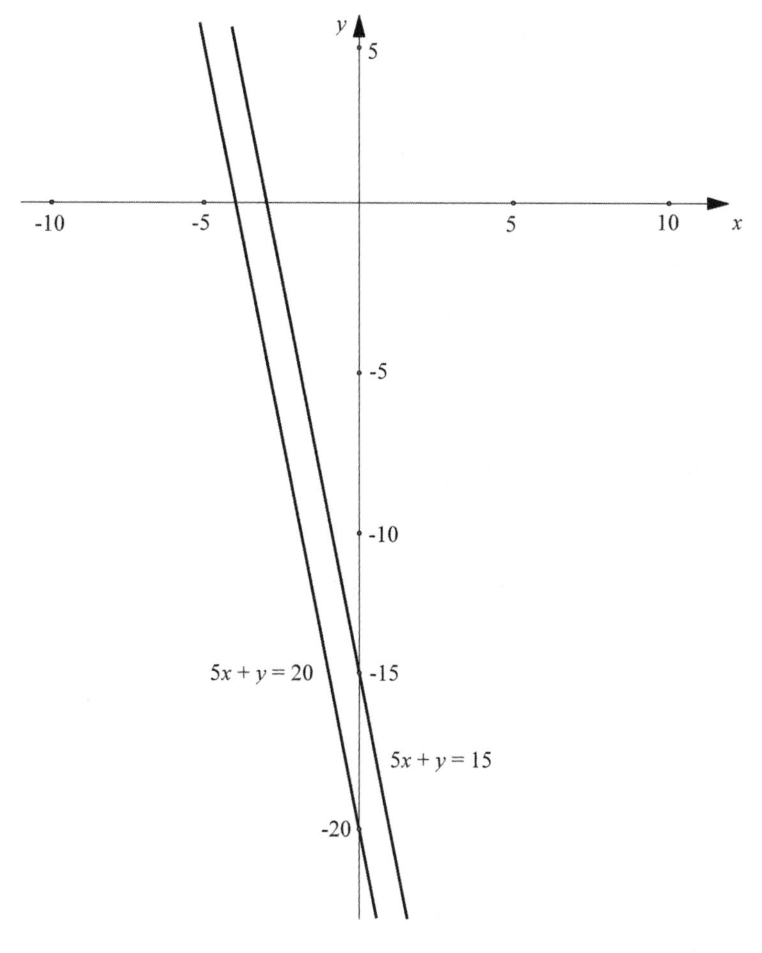

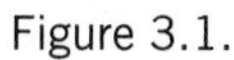
Figure 3.1.

With this reasoning we can also prove that $5 = 16$. To do this, we begin with the equation:

$$(x+1)^2 - (x+2)(x+3) = (x+4)(x+5) - (x+6)^2.$$

Then doing the indicated multiplications, we get:

$$x^2 + 2x + 1 - (x^2 + 5x + 6) = x^2 + 9x + 20 - (x^2 + 12x + 36).$$

By combining like terms, we then get: $-3x - 5 = -3x - 16$.

Adding $3x$ to both sides gives us $-5 = -16$

Lastly, multiplying by -1, we get the absurd result that $5 = 16$. From a given equation the mathematical mistake is less obvious, but as you approach the absurd result, you will notice the similarity to the previous oversight. You then ask yourself, where is the mistake? We multiplied and added correctly. The mistake lies in the original equation, which only becomes clear at the end of this discussion. In other words, assuming the existence of a solution (but there is none) of this last equation, we find the contradiction $5 = 16$. From a contradiction we conclude that one of our assumptions or some reasoning must be wrong. But here all steps are correct except our assumption that the equation has a solution. Therefore, we can conclude that our original equation has no solution.

SIMULTANEOUS EQUATIONS LEADING TO A STRANGE RESULT THROUGH AN ALGEBRAIC MISTAKE

We begin with the simultaneous equations given below:

$$\frac{x}{y} + \frac{y}{x} = 2 \qquad (1)$$

$$x - y = 4 \qquad (2)$$

Then by multiplying equation (1) by xy, we arrive at: $x^2 + y^2 = 2xy$, which can then be reworked to give us $x^2 - 2xy + y^2 = 0$.

This can be written as $(x - y)^2 = 0$, whereupon $x - y = 0$, or put another way, $x = y$.

If we substitute this value of y in equation (2), we end up with the ridiculous result $x - x = 0 = 4$, which then results in $0 = 4$. So where is the mistake?

From the given, we would assume that x and y are not zero. Actually, this system of equations has no solution. If we look at this graphically, we will notice that we have two parallel lines. With no intersection of the two lines, there can be no common solution (see figure 3.2).

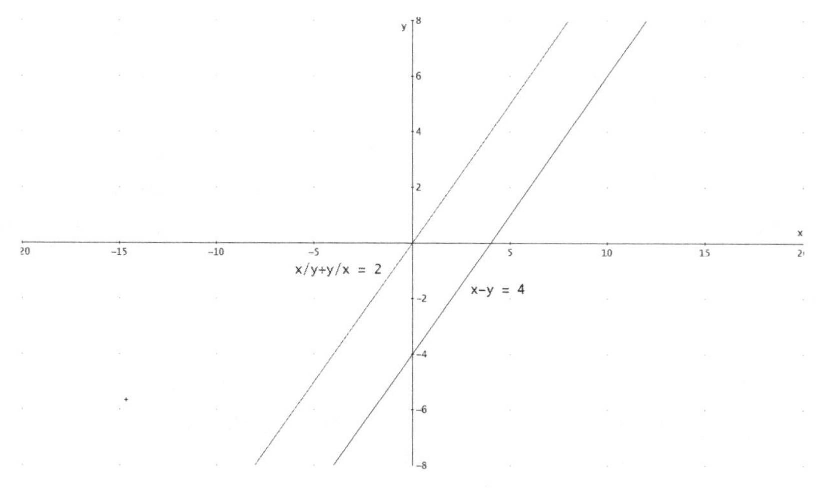

Figure 3.2.

A MISTAKE BASED ON FAULTY SQUARE-ROOT EXTRACTION LEADING AN ABSURD CONCLUSION

In this case, just follow along with the steps shown below:

$$2 = 2$$

$$3 - 1 = 6 - 4$$

$$1 - 3 = 4 - 6$$

$$1 - 3 + \frac{9}{4} = 4 - 6 + \frac{9}{4}$$

$$1 - 2 \cdot \frac{3}{2} + \frac{9}{4} = 4 - 4 \cdot \frac{3}{2} + \frac{9}{4}$$

$$\left(1 - \frac{3}{2}\right)^2 = \left(2 - \frac{3}{2}\right)^2$$

$$1 - \frac{3}{2} = 2 - \frac{3}{2}$$

$$1 = 2$$

So where is the error?

It is hidden in the step where we took the square root of both sides of the following:

$$\left(1-\frac{3}{2}\right)^2=\left(2-\frac{3}{2}\right)^2,$$

and arrived at this expression $1-\frac{3}{2}=2-\frac{3}{2}$, which ignored the negatives that need to be considered. We should have gotten the absolute values as follows:

$$\left(1-\frac{3}{2}\right)^2=\left(2-\frac{3}{2}\right)^2,$$

which would have led us to $\left|-\frac{1}{2}\right|=\left|\frac{1}{2}\right|$, which results in something that is certainly reasonable (and correct), $\frac{1}{2}=\frac{1}{2}$. Ignoring the proper square-root extraction can lead to countless mistakenly silly results.

For example, suppose we begin with $-20=-20$, and write it as $16-36=25-45$. If we add $\frac{81}{4}$ to both sides, we get: $16-36+\frac{81}{4}=25-45+\frac{81}{4}$, which is equivalent to:

$$\left(4-\frac{9}{2}\right)^2=\left(5-\frac{9}{2}\right)^2.$$

Now, taking the square root of both sides (although incorrectly, as stated above), we get the following $4-\frac{9}{2}=5-\frac{9}{2}$, which then results in the ridiculous result $4=5$.

However, had we done the work correctly by considering that the square root should result in an absolute-value statement, we would have gotten the following: $\left|4-\frac{9}{2}\right|=\left|5-\frac{9}{2}\right|$, which leads to a sensible result: $\frac{1}{2}=\frac{1}{2}$.

A SUBTLE MISTAKE IN SOLVING AN EQUATION WILL CAUSE AN ERROR

We are asked to solve the equation: $3x-\sqrt{2x-4}=4x-6$ (where $x\in\mathbf{R}$, $x\geq 2$). The usual way to solve this equation is to isolate the radical term and then square both sides as follows:

$$\sqrt{2x-4}=-(x-6)$$
$$2x-4=(-(x-6))^2=(x-6)^2$$

Simplifying this, we get $x^2 - 14x + 40 = 0$. This then produces two roots, $x_1 = 10$ and $x_2 = 4$. By substituting these two values into the original equation, we find that only x_2 is a solution, and not x_1. Why is x_1 not a valid solution? The mistake we made was to consider taking the square root of both sides, which is not an equivalent change. In other words, from $T_1 = T_2$ it follows that $T_1^2 = T_2^2$. However, the reverse is not true—as we have just experienced. In figure 3.3 we graph each side of the original equation and can see where the two equations intersect.

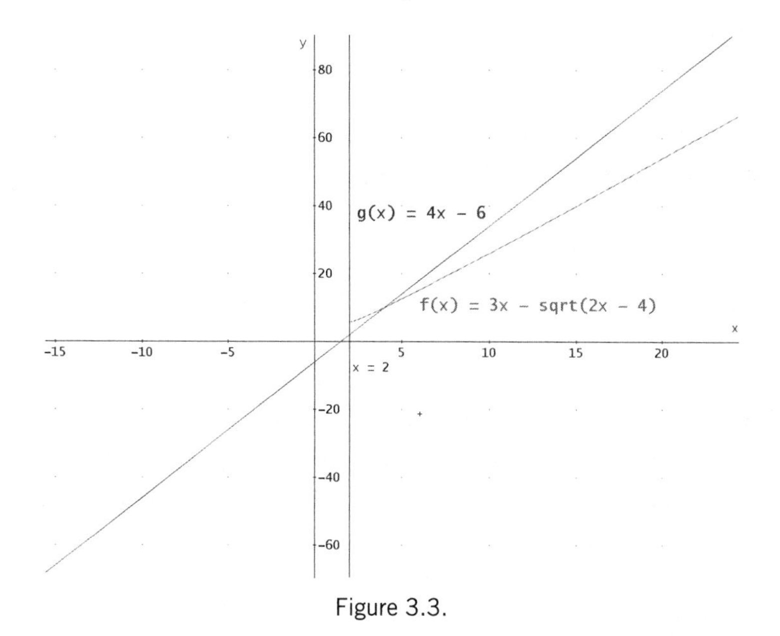

Figure 3.3.

TO "PROVE" THAT 0 = 100: STEMMING FROM A MISTAKE WITH SQUARE-ROOT EXTRACTION

We begin by letting $y = 100$, and $z = 0$. Suppose we have $x = \frac{y+z}{2}$. Then $2x = y + z$.

By multiplying both sides by $y - z$, we get $2x(y - z) = (y + z)(y - z)$.

Now, multiplying as indicated: $2xy - 2xz = y^2 - z^2$.

Rearranging terms: $z^2 - 2xz = y^2 - 2xy$.

Then adding x^2 to both sides of the equation: $z^2 - 2xz + x^2 = y^2 - 2xy + x^2$.

Factoring each of the sides of the equation: $(z - x)^2 = (y - x)^2$.

Taking the square root of each side gives us $z - x = y - x$.

This leaves us with $z = y$, or by substituting the original values for y and z, we have $0 = 100$.

Once again, the mistake lies in the oversight resulting from assuming the square root of a number is positive, when it could also be negative. So that in the next-to-last step we could have gotten $z - x = -(y - x)$, which then gets us back to what we began with, $x = \frac{y+z}{2}$.

TO SHOW THAT IF $a \neq b$, THEN $a = b$: ANOTHER MISTAKE STEMMING FROM SQUARE-ROOT EXTRACTION

There are times when our mistake could be of the same nature as that above, but well hidden and therefore easily overlooked, as in the following example.

We begin with $a \neq b$, and assume (without loss of generality) $a < b$. Further let $c = \frac{a+b}{2}$. That means that the equation is $a + b = 2c$. By multiplying both sides by $a - b$, we get $a^2 - b^2 = 2ac - 2bc$.

Then, adding $b^2 - 2ac + c^2$ to both sides of the previous equation, we get $a^2 - 2ac + c^2 = b^2 - 2bc + c^2$. Each of the two sides of this equation is a perfect square, and, thus, the equation can be written as $(a - c)^2 = (b - c)^2$. Taking the square root of both sides, $\sqrt{(a-c)^2} = \sqrt{(b-c)^2}$, gives us $a - c = b - c$, or $a = b$. Recall that we began by stating that $a \neq b$. There must be a mistake somewhere in our work. We seem to have done every step of this algebraic process correctly. However, in the last step, where we took the square root of both sides, we neglected to consider the negative values. Had we taken the result of extracting the square root of both sides of $\sqrt{(a-c)^2} = \sqrt{(b-c)^2}$ to get $a - c = -(b - c)$, we would have gotten $a - c = -b + c$, which is our original equation, $a + b = 2c$.

CAUTION MUST BE TAKEN IN SOLVING EQUATIONS, OR ELSE A MISTAKE MAY BE ENCOUNTERED

Another error arising from a surprise mistake is one very subtly hidden in the procedure of solving the equation $1 + \sqrt{x+2} = 1 - \sqrt{12-x}$.

We begin our solution to this equation by adding –1 to both sides and then squaring the two sides. This yields $x + 2 = 12 - x$, which results in $x = 5$.

If we substitute this value of x into the original equation, we get $1 + \sqrt{5+2} = 1 - \sqrt{12-5}$; and then, adding –1 to both sides and then squaring both sides of $\sqrt{5+2} = -\sqrt{12-5}$, we get $7 = 7$. This would have us think that the value of x is the correct value. It is not! If we substitute 5 in place of x in the original equation, we would have $1 + \sqrt{7} = 1 - \sqrt{7}$, which is not correct. There is no answer to this equation.

Where, then, has the mistake been made? When taking the square root, we must take the positive *and* the negative into account. We violated that rule in this process!

It is good to remember that squaring both sides of an equation is not an equivalence transformation. It yields a new equation with possibly more solutions than the given equation. Therefore, not every solution of the "squared equation" is a solution of the original equation. This is a very important rule, very often not explicitly expressed. This is when mistakes appear.

Consider the equation $x + 5 - \sqrt{x+5} = 6$. This can be written as $x - 1 = \sqrt{x+5}$. Squaring both sides and solving, we get $x^2 - 2x + 1 = x + 5$, simplified as $x^2 - 3x - 4 = 0$

Then $x = 4$ or $x = -1$. However, whereas $x = 4$ is a solution, $x = -1$ is not a solution. This is a typical mistake made in algebra classes. Again, the square-root process did not take the negative into account.

This absurdity can be taken further, for if we want to prove that $5 = 1$, we then subtract 3 from both sides to get $2 = -2$, and then squaring both sides, we get $4 = 4$. Therefore, 5 must have equaled 1!

A MISTAKE WITH POWERS CAN LEAD YOU AWRY

Imagine someone solving the equation $\left(\frac{2}{3}\right)^x = \left(\frac{3}{2}\right)^3$ as follows:

Begin with the given $\left(\frac{2}{3}\right)^x = \left(\frac{3}{2}\right)^3$ and then apply the powers to the fractions to get $\frac{2^x}{3^x} = \frac{3^3}{2^3}$ Multiplying by the common denominator, $2^3 \cdot 3^x$ leaves us with $2^3 \cdot 2^x = 3^3 \cdot 3^x$.

Following the rules of exponents, we have: $2^{3+x} = 3^{3+x}$.

If two equal powers have equal exponents, we might conclude that the bases must be equal as well, therefore, $2 = 3$. Something is not correct here. Where might the mistake be? The last step is wrong! The correct solution is $x = -3$, which makes each of the two sides equal to 1.

You may want to see what this looks like graphically; so we provide the two functions $f(x) = 2^{3+x}$ and $g(x) = 3^{3+x}$, and show the graph in figure 3.4.

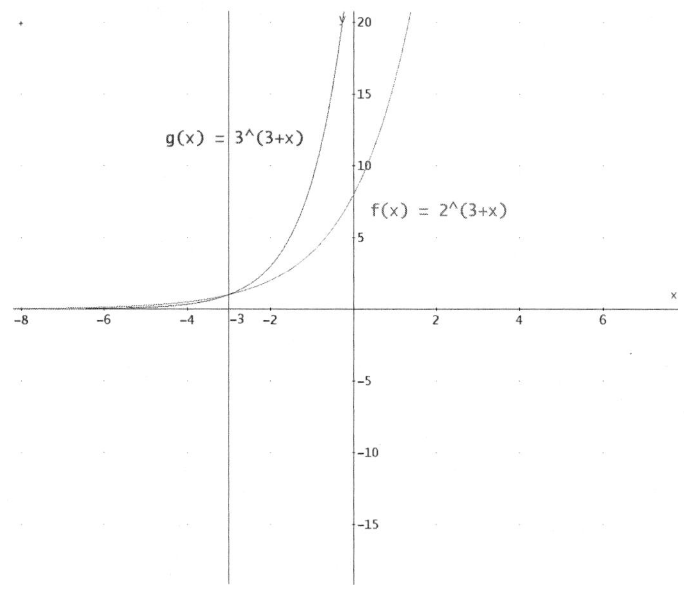

Figure 3.4.

A SUBTLE OVERSIGHT, OR MISTAKE, WITH THE BINOMIAL THEOREM

We know that $(a + b)^2 = a^2 + 2ab + b^2$. This formula is an application of the binomial theorem, which provides a formula for obtaining the value of a binomial raised to *any* positive integer power n. The formula is:

$$(a + b)^n = a^n + na^{n-1}b + \frac{n(n-1)}{2!}a^{n-2}b^2 + \ldots + \frac{n(n-1)}{2!}a^2b^{n-2} + nab^{n-1} + b^n.$$

If $n = 2$, we get $(a + b)^2 = a^2 + 2ab + b^2$.

If $n = 1$, we get $(a + b)^1 = a^2 + b^1 = a + b$.

Using the formula of the binomial theorem, we can also be led by mistake to a weird conclusion.

When $n = 0$, we get $1 = 1 + 0 + 0 + 0 + \ldots + 0 + 0 + 0 + 1$, or $1 = 2$.

Because when $n = 0$, then $(a + b)^0 = 1$, $a^0 = 1$, $b^0 = 1$.

That we reached an absurd result would have us believe that a mistake was made, and indeed one was.

But where is the error?

$(a + b)^n = (a + b)^0 = 1$—that is OK.

Now look at the right side of the original equation more precisely, for therein lies the reason for our mistake.

On the right side of the equation there is for $n = 0$ only *one* term (not *two or more* terms), which we produced by substituting into the binomial formula. That term is $\frac{1}{0!}a^{0-0}b^0 = 1 \cdot a^0 \cdot b^0 = 1 \cdot 1 \cdot 1 = 1$

We can also see that on the Pascal triangle, which can be used to determine the coefficients of the terms of the binomial expansion as follows, and as shown in figure 3.5:

$$(a + b)^0 = 1$$

$$(a + b)^1 = 1a + 1b$$

$$(a + b)^2 = 1a^2 + 2ab + 1b^2$$

$$(a + b)^3 = 1a^3 + 3a^2b + 3ab^2 + 1b^3$$

$$\ldots$$

$$(a + b)^n = \frac{1}{0!}a^n + \frac{n}{1!}a^{n-1}b + \frac{n(n-1)}{2!}a^{n-2}b^2 + \ldots + \frac{n(n-1)}{2!}a^2b^{n-2} + \frac{n}{1!}ab^{n-1} + \frac{1}{0!}b^n$$

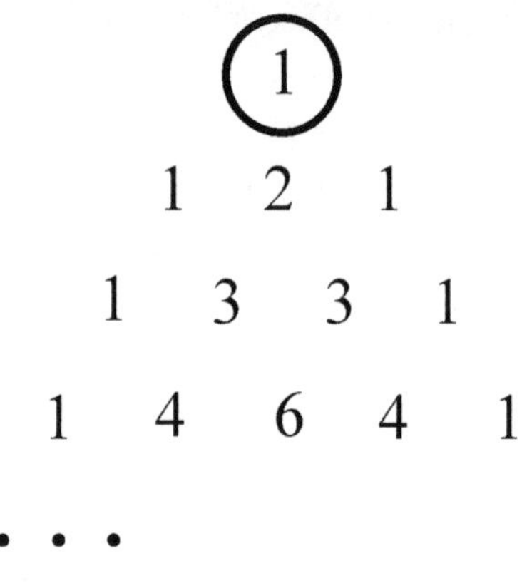

Figure 3.5.

But be careful—if $a = b = 0$, you could have a problem! What is the value of 0^0? This is an expression that we leave undefined, and many calculators do not react to this input, yet some deliver the value 1, since "anything" taken to the zero power seems to be 1.

As we said at the outset, the binomial theorem only holds for *positive* values of n. Notice the mistake that using zero has caused. That is why we specify the value of n—to avoid such ridiculous results!

TO SHOW THAT IF p IS POSITIVE, IT IS ACTUALLY NEGATIVE: A MISTAKE WITH INEQUALITIES

We begin by assuming that both p and q are positive, and we are going to show that p is negative. Clearly, the inequality $2q - 1 < 2q$ is a true statement. Suppose we multiply both sides by $-p$ to get $-2pq + p < -2pq$. If we then add $-2pq$ to both sides of this inequality, we end up with $p < 0$, which is to say, p is negative. How can this be, when we started off with a positive p? Where was the mistake?

We violated a rule for inequalities. That is, *when multiplying (or dividing) both sides of an inequality by a negative number, the inequality symbol must be reversed.*

Look at a simple example: $2 < 3$, but when we multiply both sides of the inequality by -1, we have the following $2 \cdot (-1) = -2$, which is greater than $3 \cdot (-1) = -3$, or simply written $-2 > -3$.

We can see this mistake played out in a less obvious way in the following example.

TO SHOW THAT ANY POSITIVE NUMBER IS GREATER THAN ITSELF

We shall begin with the two positive numbers p and q, where $p > q$. We will now multiply both sides of this inequality by q to get $pq > q^2$. Subtracting p^2 from both sides of the inequality gives us $pq - p^2 > q^2 - p^2$. By factoring both sides, we get $p(q - p) > (q + p)(q - p)$.

Dividing both sides by $(q - p)$ leaves us with $p > q + p$, which says that p is greater than itself. That's absurd! So where was the mistake made? Since $p > q$, it must follow that $(q - p)$ is negative. We made the mistake of not reversing the inequality symbol when dividing both sides of the inequality by the negative term $q - p$.

The structure of this example follows the "proof" of "1 = 2: A Mistake Based on Division by Zero"—instead of division by zero in an equation, we had a division by a negative term for an inequality. This absurd result can be taken a step further. With the given $p > q$ and our newly found result $p > q + p$, we can add the two inequalities to get $2p > 2q + p$. Subtracting p from both sides, we get $p > 2q$. So if having $p > q$, and $p > 2q$, we can conclude that $p > 2q$. Similar reasoning would allow us to then conclude that $p > 4q$. This can continue in the same way to further absurdities.

To show that $\frac{1}{8} > \frac{1}{4}$ is an obvious mistake, you may have to recall what $\log(x)$ is. Remember that $y = \log_b(x)$ means the same as $b^y = x$, namely, y is the power (essentially an exponent) to which a base b is raised to yield a given number x. For the base $b = 10$, we write $y = \log_{10}(x) = \log(x)$. In addition we know the rule or property that $y \cdot \log(x) = \log(x^y)$.

Beginning with the obvious $3 > 2$, and then multiplying both sides of this inequality by $\log\frac{1}{2}$, we get $3 \cdot \log\frac{1}{2} > 2 \cdot \log\frac{1}{2}$, or, using the above log rule,

$\log\left(\frac{1}{2}\right)^3 > \log\left(\frac{1}{2}\right)^2$, which leads us (by the power to base 10) to $\left(\frac{1}{2}\right)^3 > \left(\frac{1}{2}\right)^2$,

which in a more familiar form is $\frac{1}{8} > \frac{1}{4}$—clearly an absurdity!

Where did we make the mistake? This was a very well-camouflaged situation where we needed to note that $\log\frac{1}{2}$ is a negative quantity, and it would have required us to reverse the inequality symbol (i.e., we could ask ourselves: "10 to which power is $\frac{1}{2}$?" A negative number will surely arise).

This was admittedly a more difficult-to-find violation of the rule that multiplication of an inequality by a negative quantity requires reversing the inequality sign.

While on the topic of logs, consider the following mistake.

ANOTHER LOG MISTAKE THAT TELLS US THAT 1 = –1

We know that $(-1)^2 = 1$. We can then inspect the log of each of these numbers—which ought to be equal:

$\log(-1)^2 = \log 1$. Then $2 \cdot \log(-1) = \log 1$. However, $\log 1 = 0$, since any number to the zero power is 1.

Therefore, $2 \cdot \log(-1)$ must also equal 0. Then $\log(-1) = 0$.

Since we now showed that both of these are equal to zero, we can say that because $\log(-1) = \log 1$, it then must follow that $-1 = 1$. Do all numbers—positive and negative—have logarithms? Therein lies the mistake! Logarithms of negative numbers are not defined for real numbers.

A COMMON MISTAKE WITH LOGARITHMS

We wish to solve the equation $2^x = 128$ in the following way:

First we take the log of both sides: $\log 2^x = \log 128$. Using the rule of logs, we get: $x \cdot \log 2 = \log 128$. Then we divide both sides of the equation by log 2:

$$x = \frac{\log 128}{\log 2}.$$

Because the log of a quotient is the difference of the logs, we have $x = \log 128 - \log 2$, which leads to $x \approx 2.107209969 - 0.3010299956 = 1.806179973$.

However, this is not the correct answer, since the correct value of $\frac{\log 128}{\log 2} = 7$, and not $x \approx 1.806179973$.

Where, then, is the mistake? We had no right to not divide the logs when we were presented with the expression $\frac{\log 128}{\log 2}$, since it is *not* equal to $\log 128 - \log 2$.

CAUTION MUST BE EXERCISED WHEN WORKING WITH INEQUALITIES TO AVOID A MISTAKE

Consider the following inequality:

$$\left(\frac{1}{6}\right)^n \leq 0.01,$$

where we seek to find the natural numbers n to make this true. We can begin by taking the log of both sides:

$$\log\left(\frac{1}{6}\right)^n \leq \log 0.01.$$

It then follows that $n \cdot \log\left(\frac{1}{6}\right)^n \leq \log 0.01$. By dividing both sides by $\log\frac{1}{6}$, we get:

$$n \leq \frac{\log 0.01}{\log\frac{1}{6}} = \frac{\log\frac{1}{100}}{\log\frac{1}{6}} = \frac{\log 1 - \log 100}{\log 1 - \log 6} = \frac{0-2}{0-\log 6} = \frac{2}{\log 6} = 2.570194417\ldots$$

This would imply that the solutions of this inequality are the natural numbers: 0, 1, and 2. However, if we substitute these values for n in the original inequality, we find that this is a mistaken solution. Where, then, has the mistake been made?

Yes, in camouflaged fashion, we divided by a negative and that requires reversing the inequality sign.

$\log\frac{1}{6} < 0$; therefore, when we divided by $\log\frac{1}{6}$, the equality symbol $\leq$ had to be reversed to $\geq$. Thus, the correct solution to the given inequality is $n \geq \frac{2}{\log 6} = 2.570194417\ldots$, so that all natural numbers n, where $n > 2$ satisfies the inequality.

EXTENDING THE DISTRIBUTIVE PROPERTY INCORRECTLY: A COMMON (YET AVOIDABLE) MISTAKE

The distributive property is one of the basic aspects of algebra. It is best demonstrated algebraically as $a(b + c) = ab + ac$, or $(a + b)c = ac + bc$. We can also show it as: $3(a + b) = 3a + 3b$.

We should also recall that although multiplication is distributive over addition, the reverse, addition over multiplication, does not work. That is, $a + (b \cdot c) \neq (a + b) \cdot (a + c)$, or $(a \cdot b) + c \neq (a + c) \cdot (b + c)$.

There are many mistakes made in algebra that should easily be avoided. Here are some of these mistakes that stem from an incorrect generalization of the distributive property: $(a + b)^2 = a^2 + b^2$, and $(a - b)^2 = a^2 - b^2$.

This mistake then gets expanded when we consider the general case $(a + b)^n = a^n + b^n$ and $(a - b)^n = a^n - b^n$.

When we let $n = \frac{1}{2}$, we find the following further mistakes:

$$\sqrt{a+b} = \sqrt{a} + \sqrt{b} \text{ and } \sqrt{a-b} = \sqrt{a} - \sqrt{b};$$

$$\sqrt[n]{a+b} = \sqrt[n]{a} + \sqrt[n]{b} \text{ and } \sqrt[n]{a-b} = \sqrt[n]{a} - \sqrt[n]{b}.$$

Furthermore, when $n = -1$ in $(a + b)^n = a^n + b^n$, or $(a - b)^n = a^n - b^n$, we get:

$$(a + b)^{-1} = a^{-1} + b^{-1} \text{, which can be written as } \frac{1}{a+b} = \frac{1}{a} + \frac{1}{b},$$

$$\text{and } (a - b)^{-1} = a^{-1} - b^{-1} \text{, which can be written as } \frac{1}{a-b} = \frac{1}{a} - \frac{1}{b}.$$

We can also find mistakes when we try to apply the distributive property to logarithms:

$$\log(a + b) = \log a + \log b, \text{ and } \log(a - b) = \log a - \log b.$$

$$\log(a \cdot b) = \log a \cdot \log b, \text{ and } \log\frac{a}{b} = \frac{\log a}{\log b}.$$

The correct distribution is: $\log(a \cdot b) = \log a + \log b$, and $\log\frac{a}{b} = \log a - \log b$.

We can also see the distributive property mistakenly applied to the absolute value function as $|a + b| = |a| + |b|$, and $|a - b| = |a| - |b|$.

We should also note that some have extended this mistaken use of the distributive property to trigonometry as seen with $\sin(\alpha + \beta) = \sin \alpha + \sin \beta$, and $\sin(\alpha - \beta) = \sin \alpha - \sin \beta$ (where α, β, are real numbers); and $\cos(\alpha + \beta) = \cos \alpha + \cos \beta$, and $\cos(\alpha - \beta) = \cos \alpha - \cos \beta$ (where α, β, are real numbers).

Such mistaken uses of the distributive property should be avoided, with the understanding that there do exist correct applications of this important property.

We can see some of these mistakes more clearly by using numbers instead of variables:

$(a + b)^2 = a^2 + b^2$ $\quad$ $(2 + 1)^2 = 3^2 = 9$, however, $2^2 + 1^2 = 4 + 1 = 5$.

$(a - b)^2 = a^2 - b^2$ $\quad$ $(2 - 1)^2 = 1^2 = 1$, however, $2^2 - 1^2 = 4 - 1 = 3$.

$(a + b)^n = a^n + b^n$ $(n = 3:)$ $\quad$ $(2 + 1)^3 = 3^3 = 27$, however, $2^3 + 1^3 = 8 + 1 = 9$.

$(a - b)^n = a^n - b^n$ $(n = 3:)$ $\quad$ $(2 - 1)^3 = 1^3 = 1$, however, $2^3 - 1^3 = 8 - 1 = 7$.

$\sqrt{a+b} = \sqrt{a} + \sqrt{b}$ $\quad$ $\sqrt{16+9} = \sqrt{25} = 5$, however, $\sqrt{16} + \sqrt{9} = 4 + 3 = 7$.

$\sqrt[n]{a+b} = \sqrt[n]{a} + \sqrt[n]{b}$ $(n = 3:)$ $\quad$ $\sqrt[3]{27+8} = \sqrt[3]{35} \approx 3.2711$, however, $\sqrt[3]{27} + \sqrt[3]{8} = 3 + 2 = 5$.

$\sqrt[n]{a-b} = \sqrt[n]{a} - \sqrt[n]{b}$ $(n = 3:)$ $\quad$ $\sqrt[3]{27-8} = \sqrt[3]{19} \approx 2.6684$, however, $\sqrt[3]{27} - \sqrt[3]{8} = 3 - 2 = 1$.

$\sqrt{a-b} = \sqrt{a} - \sqrt{b}$ $\quad$ $\sqrt{25-9} = \sqrt{16} = 4$, however, $\sqrt{25} - \sqrt{9} = 5 - 3 = 2$.

$\frac{1}{a+b} = \frac{1}{a} + \frac{1}{b}$ $\quad$ $\frac{1}{2+1} = \frac{1}{3}$, however, $\frac{1}{2} + \frac{1}{1} = \frac{3}{2}$.

$\frac{1}{a-b} = \frac{1}{a} - \frac{1}{b}$ $\quad$ $\frac{1}{2-1} = 1$, however, $\frac{1}{2} - \frac{1}{1} = -\frac{1}{2}$.

$\log(a+b) = \log a + \log b$	$\ln(e+e) = \ln 2e = \ln 2 + \ln e = \ln 2 + 1 \approx 1.693147180$, however, $\ln e + \ln e = 1 + 1 = 2$.
$\log(a-b) = \log a - \log b$	$\ln(2e-e) = \ln e = 1$, however, $\ln 2e - \ln e = \ln 2 + \ln e - \ln e = \ln 2 \approx 0.693147180$.
$\lvert a+b\rvert = \lvert a\rvert + \lvert b\rvert$	$\lvert 3+(-4)\rvert = \lvert -1\rvert = 1$, however, $\lvert 3\rvert + \lvert(-4)\rvert = 3 + 4 = 7$.
$\lvert a-b\rvert = \lvert a\rvert - \lvert b\rvert$	$\lvert 3-(-4)\rvert = \lvert 7\rvert = 7$, however, $\lvert 3\rvert - \lvert -4\rvert = 3 - 4 = -1$.
$\sin(\alpha+\beta) = \sin\alpha + \sin\beta$	$\sin(\frac{\pi}{3}+\frac{\pi}{6}) = \sin\frac{\pi}{2} = 1$, however, $\sin\frac{\pi}{3} + \sin\frac{\pi}{6} = \frac{\sqrt{3}}{2} + \frac{1}{2} = \frac{\sqrt{3}+1}{2}$.
$\sin(\alpha-\beta) = \sin\alpha - \sin\beta$	$\sin(\frac{\pi}{3}-\frac{\pi}{6}) = \sin\frac{\pi}{6} = \frac{1}{2}$, however, $\sin\frac{\pi}{3} - \sin\frac{\pi}{6} = \frac{\sqrt{3}}{2} - \frac{1}{2} = \frac{\sqrt{3}-1}{2}$.

MISTAKEN UNDERSTANDINGS OF INFINITY LEADS TO ABSURD RESULTS

There are lots of curiosities that arise from our lack of true understanding of the concept of infinity. For example, many folks have difficulty understanding that the set of all natural numbers {0, 1, 2, 3, 4, . . .} has the same number of members as the set of the even integers {0, 2, 4, 6, 8, . . .}, which is missing all the odd numbers that are included in the first set. How, then, can the two sets be of equal size? One way of comparing the size of the sets is to match them member for member. Therefore, we see that for every member of the set of natural numbers there will always be a member of the set of even integers, namely, one twice as large, and vice versa, for every member of the set of even integers there will always be a member of the set of natural numbers, that is, one half as large. Therefore, they must have "the same number" of members in the two sets. This may be counterintuitive, but it is true. This is just one of the considerations of the concept of infinity.

We can also misuse the concept of infinity—that is, using it mistakenly —as shown in the following examples.

TO SHOW THAT 1 = 0 WITH A MISTAKEN UNDERSTANDING OF INFINITY

Let's begin with $S = 1 - 1 + 1 - 1 + 1 - 1 + \ldots$

We can group these numbers in the series as follows:

$$S = (1 - 1) + (1 - 1) + (1 - 1) + (1 - 1) + \ldots$$
$$S = 0 + 0 + 0 + 0 + \ldots$$
$$S = 0.$$

We can also group these numbers as follows:

$$S = 1 - (1 - 1) - (1 - 1) - (1 - 1) - (1 - 1) - \ldots$$
$$S = 1 - 0 - 0 - 0 - 0 - \ldots$$
$$S = 1.$$

From these two results for the value of S, it would be expected to conclude that $1 = 0$, since both equal S. The reason for this mistake rests in the convergency of this series. An *absolutely convergent* series is one where the series converges to a definite value at infinity, but also if the series formed by changing all the plus signs to minus signs converges as well. A *conditionaly convergent* series—one that is not *absolutely convergent*—was proven by the German mathematician Bernhard Riemann (1826–1866) in 1854 to be one where the terms could be rearranged so that the limit of the series could be to equal almost any number. Every convergent, but not absolutely convergent, series can be rearranged so that it is convergent to *every* given real number. Therefore, we end up with weird results by rearranging the terms of the conditionally convergent series.

We can also demonstrate this paradox when the number 1 is replaced by the number n: $(n - n) - (n - n) - \ldots = 0$, but $n - (n - n) - (n - n) - \ldots = n$.

This paradox has long been known as it was first published by the mathematician Bernard Bolzano (1781–1848) in his book *Paradoxien des*

Unendlichen (Leipzig: Meiner & Reclam, 1851; reprinted 1975). Bear in mind that when dealing with infinite series (convergent or not), parentheses must be used very cautiously.

MISTAKES WITH INFINITE SERIES THAT LEAD TO SHOWING THAT 2 = 3

When we take the reciprocal of each member of an arithmetic series, we form what is called a harmonic series. Consider the following harmonic series, which appears to steadily grow larger as more terms are added:

$$\frac{1}{1}+\frac{1}{2}+\frac{1}{3}+\frac{1}{4}+\frac{1}{5}+\frac{1}{6}+\frac{1}{7}+\frac{1}{8}+\frac{1}{9}+\frac{1}{10}+\ldots.$$

Such a series is called a divergent series. If we partition this series as follows,

$$\frac{1}{1}+\frac{1}{2}+\left(\frac{1}{3}+\frac{1}{4}\right)+\left(\frac{1}{5}+\frac{1}{6}+\frac{1}{7}+\frac{1}{8}\right)+$$

$$\left(\frac{1}{9}+\frac{1}{10}+\frac{1}{11}+\frac{1}{12}+\frac{1}{13}+\frac{1}{14}+\frac{1}{15}+\frac{1}{16}\right)+\left(\frac{1}{17}+\ldots,\right.$$

we notice that the sum of the terms in each of the parentheses is greater than $\frac{1}{2}$.

Suppose we now alternate the signs in this harmonic series:

$$\frac{1}{2}-\frac{1}{2}+\frac{1}{3}-\frac{1}{4}+\frac{1}{5}-\frac{1}{6}+\frac{1}{7}-\frac{1}{8}+\frac{1}{9}-\frac{1}{10}\pm\ldots.$$

To get the value of s (below) we will apply the commutative and associative laws to rearrange the terms as follows:

$$s=\frac{1}{1}-\frac{1}{2}-\frac{1}{4}+\frac{1}{3}-\frac{1}{6}-\frac{1}{8}+\frac{1}{5}-\frac{1}{10}-\frac{1}{12}\pm\ldots$$

$$=\left(1-\frac{1}{2}\right)-\frac{1}{4}+\left(\frac{1}{3}-\frac{1}{6}\right)-\frac{1}{8}+\left(\frac{1}{5}-\frac{1}{10}\right)-\frac{1}{12}\pm\ldots$$

$$=\frac{1}{2}-\frac{1}{4}+\frac{1}{6}-\frac{1}{8}+\frac{1}{10}-\frac{1}{12}\pm\ldots$$

$$=\frac{1}{2}\cdot\left(\frac{1}{1}-\frac{1}{2}+\frac{1}{3}-\frac{1}{4}+\frac{1}{5}-\frac{1}{6}+\frac{1}{7}-\frac{1}{8}+\frac{1}{9}-\frac{1}{10}\pm\ldots\right)=\frac{1}{2}\cdot s.$$

Strangely enough, we now have the curious result that $s = \frac{1}{2} \cdot s$, which implies that $s = 0$.

This is quite noteworthy, since if we were to partition the series as:

$$s = \frac{1}{1} - \frac{1}{2} + \frac{1}{3} - \frac{1}{4} + \frac{1}{5} - \frac{1}{6} + \frac{1}{7} - \frac{1}{8} + \frac{1}{9} - \frac{1}{10} \pm \dots$$
$$= \left(1 - \frac{1}{2}\right) + \left(\frac{1}{3} - \frac{1}{4}\right) + \left(\frac{1}{5} - \frac{1}{6}\right) + \dots,$$

then we have each parenthetical expression greater that zero. This leads to the contradictory result that $s > 0$.

So, then, where is the mistake? Well, the correct answer[1] is: $s = \ln 2$, which is the natural logarithm[2] of 2.

We can further complicate the situation if we rearrange the terms of the series as follows—without adding or deleting any member of the series:

$$\frac{1}{1} + \frac{1}{3} - \frac{1}{2} + \frac{1}{5} + \frac{1}{7} - \frac{1}{4} + \frac{1}{9} + \frac{1}{11} - \frac{1}{6} \pm \dots.$$

In this case, the series takes on a sum of $\frac{3}{2} \cdot \ln 2$. This would imply that $\ln 2 = \frac{3}{2} \cdot \ln 2$, which is to say that $1 = \frac{3}{2}$, or $2 = 3$.

The reason for the various sums of this logarithmic series is that it is not *absolutely* convergent, but *conditionally* convergent.

MISTAKES WITH SERIES WILL LEAD US TO SHOW THAT –1 IS ACTUALLY POSITIVE

We have other series-related mistakes that merit viewing. Consider the following.

We begin with the series: $S = 1 + 2 + 4 + 8 + 16 + 32 + 64 + \dots$.

This, of course, appears to tell us that S is positive.

We now multiply both sides by 2 to get:

$$2S = 2 + 4 + 8 + 16 + 32 + 64 + 128 + \dots$$
$$= (-1 + 1) + 2 + 4 + 8 + 16 + 32 + 64 + 128 + \dots$$
$$= -1 + (1 + 2 + 4 + 8 + 16 + 32 + 64 + 128 + \dots)$$
$$= -1 + S, \text{ which is } 2S = S - 1. \text{ Therefore } S = -1.$$

We begin with a positive value for S, and then end up with a negative value for S by a misuse of the parentheses.[3]

This paradox was already known to the famous Swiss mathematician Jakob I. Bernoulli (1655–1705).

MISTAKES WITH SERIES TO SHOW THAT ZERO IS POSITIVE

We begin with the following values for m and n:

$$m = \frac{1}{1} + \frac{1}{3} + \frac{1}{5} + \frac{1}{7} + \frac{1}{9} + \ldots$$

$$n = \frac{1}{2} + \frac{1}{4} + \frac{1}{6} + \frac{1}{8} + \frac{1}{10} + \ldots$$

$$2n = \frac{2}{2} + \frac{2}{4} + \frac{2}{6} + \frac{2}{8} + \frac{2}{10} + \ldots$$

$$= \frac{1}{1} + \frac{1}{2} + \frac{1}{3} + \frac{1}{4} + \frac{1}{5} + \frac{1}{6} + \frac{1}{7} + \frac{1}{8} + \frac{1}{9} + \frac{1}{10} + \ldots.$$

Therefore, $2n = m + n$, so that $m = n$ or $0 = m - n$.

However, if we subtract corresponding terms, we get:

$$m - n = \left(\frac{1}{1} - \frac{1}{2}\right) + \left(\frac{1}{3} - \frac{1}{4}\right) + \left(\frac{1}{5} - \frac{1}{6}\right) + \left(\frac{1}{7} - \frac{1}{8}\right) + \left(\frac{1}{9} - \frac{1}{10}\right) + \ldots.$$

Since each parenthetical term is positive, $m - n$ must also be positive. Considering the above, we now have a dilemma: How is zero is positive? Where does the mistake lie? You know the answer—it is the same as our earlier paradoxes.

MISTAKEN WORK WITH SERIES LEADS TO THE CONCLUSION THAT $\infty = -1$

Suppose we were to add the terms $1 + 1 + 1 + 1 + 1 + \ldots$ to infinity, and then write them as: $(-1 + 2) + (-2 + 3) + (-3 + 4) + (-4 + 5) + \ldots$. Then the sum would surely be infinity (∞).

However, we could write this sum by arranging the parentheses differently, as:

$$-1 + (2 + (-2)) + (3 + (-3)) + (4 + (-4)) + \ldots = -1 + 0 + 0 + 0 + \ldots = -1.$$

Does this then imply that $-1 = \infty$? Where is the error?

Our intuition works only as long as we can count the numbers. Here we have infinitely many numbers, so our intuition could be misleading. Interestingly, in certain branches of advanced mathematics, researchers have found it productive to use these "flawed" definitions of infinite series and see what happens when we let them be true. Andrew Wiles (1953–) used properties of this alternative system to prove the famous "Fermat's last theorem"!

MISTAKEN WORK WITH A SERIES CAN RESULT IN 0 = 1

We shall begin by considering the following series: $\frac{1}{1\cdot 2} + \frac{1}{2\cdot 3} + \frac{1}{3\cdot 4} + \frac{1}{4\cdot 5} + \ldots$, which we will show has a value of 1. Follow along as we now evaluate the sum of the series:

$$\frac{1}{1\cdot 2} + \frac{1}{2\cdot 3} + \frac{1}{3\cdot 4} + \frac{1}{4\cdot 5} + \ldots = \frac{1}{2} + \frac{1}{6} + \frac{1}{12} + \frac{1}{20} + \ldots$$

$$= \left(\frac{1}{1} - \frac{1}{2}\right) + \left(\frac{1}{2} - \frac{1}{3}\right) + \left(\frac{1}{3} - \frac{1}{4}\right) + \left(\frac{1}{4} - \frac{1}{5}\right) + \ldots$$

$$= \frac{1}{1} + \left(-\frac{1}{2} + \frac{1}{2}\right) + \left(-\frac{1}{3} + \frac{1}{3}\right) + \left(-\frac{1}{4} + \frac{1}{4}\right) + \left(-\frac{1}{5} + \frac{1}{5}\right) + \ldots.$$

Since each parenthetical expression has the value of zero, the sum of the series is simply 1:

$$1 = \frac{1}{1\cdot 2} + \frac{1}{2\cdot 3} + \frac{1}{3\cdot 4} + \frac{1}{4\cdot 5} + \frac{1}{5\cdot 6} + \frac{1}{6\cdot 7} + \ldots.$$

If we now subtract the first term, $\frac{1}{2}$, from both sides of this equation, we get:

$$\frac{1}{2} = \frac{1}{2\cdot 3} + \frac{1}{3\cdot 4} + \frac{1}{4\cdot 5} + \frac{1}{5\cdot 6} + \frac{1}{6\cdot 7} + \ldots.$$

Subtracting the first term, $\frac{1}{2\cdot 3}$, from both sides of this equation, we get:

$$\frac{1}{3} = \frac{1}{3\cdot 4} + \frac{1}{4\cdot 5} + \frac{1}{5\cdot 6} + \frac{1}{6\cdot 7} + \ldots.$$

Continuing this process gives us:

$$\frac{1}{4} = \frac{1}{4\cdot 5} + \frac{1}{5\cdot 6} + \frac{1}{6\cdot 7} + \ldots.$$

Now taking the sums of the left sides of each of these equations, we end up with $1 + \frac{1}{2} + \frac{1}{3} + \frac{1}{4} + \ldots$, while on the right sides we get the following sum:

$$= \left(\frac{1}{1\cdot 2} + \frac{1}{2\cdot 3} + \frac{1}{3\cdot 4} + \frac{1}{4\cdot 5} + \frac{1}{5\cdot 6} + \frac{1}{6\cdot 7} + \ldots\right)$$

$$+ \left(\frac{1}{2\cdot 3} + \frac{1}{3\cdot 4} + \frac{1}{4\cdot 5} + \frac{1}{5\cdot 6} + \frac{1}{6\cdot 7} + \ldots\right)$$

$$+ \left(\frac{1}{3\cdot 4} + \frac{1}{4\cdot 5} + \frac{1}{5\cdot 6} + \frac{1}{6\cdot 7} + \ldots\right) + \left(\frac{1}{4\cdot 5} + \frac{1}{5\cdot 6} + \frac{1}{6\cdot 7} + \ldots\right) + \ldots$$

$$= \frac{1}{1\cdot 2} + 2\cdot\frac{1}{2\cdot 3} + 3\cdot\frac{1}{3\cdot 4} + 4\cdot\frac{1}{4\cdot 5} + \ldots$$

$$= \frac{1}{2} + \frac{1}{3} + \frac{1}{4} + \ldots.$$

Then, when we equate it to the above right-side sum, we get:

$$1 + \frac{1}{2} + \frac{1}{3} + \frac{1}{4} + \ldots = \frac{1}{2} + \frac{1}{3} + \frac{1}{4} + \ldots,$$

which essentially gives us the ridiculous result of $1 = 0$.

Again, we ask, where is the error in the above calculation? The answer rests with the difference between a divergent versus convergent series.

WORKING WITH INFINITY CAUSES CONFUSION AND CAN LEAD TO $0 = \infty$

A magician places a coin in a box. After half an hour, he removes the coin from the box and places a second, third, and fourth coin in the box. A quarter of an hour later, the magician takes out one coin and replaces it with three new coins. After an eighth of an hour, magician takes out another coin and replaces it with three new coins. He continues this process for one hour. How many coins will there be in the box at the end of the hour?

Solution 1:

At every step, one coin is removed and three new ones are added to the box. This means that at every step, two coins are added to the collection in the box. Therefore, after n steps there will be $2n + 1$ coins in the box. With this procedure, the magician will have completed an infinite number of steps in one hour. Therefore, there should be an infinite number of coins in the box.

Solution 2:

Which would be the last coin to be in the box? Every coin that was put in the box could be assigned a number. Suppose n is the number of one of the coins in the box. Then at the nth step of the procedure, this coin will then be removed from the box. That would imply that every coin placed in the box will at one point in time be removed from the box. Consequently, at the end of the hour there will be no coins in the box.

This implies the following $0 = \infty$. Where, then, does the mistake lie? Recall the peculiarities of infinite series as a clue.

There are also mistakes made by assuming that imaginary numbers—that is, numbers that result from taking the square root of a negative number—can be treated the same way as real numbers. The following example will show the mistake of treating these numbers incorrectly.

MISTAKEN APPLICATION OF COMPLEX NUMBERS TO PROVE THAT −1 = +1

We begin with the product of two imaginary numbers $\sqrt{-1}$, and apply the rules we know of real numbers:

$$\sqrt{-1} \cdot \sqrt{-1} = \sqrt{(-1) \cdot (-1)} = \sqrt{+1} = 1.$$

This time we will evaluate the product as follows:

$$\sqrt{-1} \cdot \sqrt{-1} = \left(\sqrt{-1}\right)^2 = -1.$$

Therefore, since the given product provides us with two values, we would have to conclude that $-1 = +1$.

Something must be wrong, since clearly -1 is not equal to $+1$. This mistake is one that depends on a definition in mathematics—one that we may even think arose to avoid this dilemma. That is, that the product $\sqrt{a} \cdot \sqrt{b} = \sqrt{a \cdot b}$ does not hold true when a and b are negative. Therefore, $\sqrt{-1} \cdot \sqrt{-1} = \sqrt{(-1) \cdot (-1)}$ is wrong! However, it is true that $\sqrt{(-1) \cdot (-1)} = \sqrt{1}$.

Analogously, a similar dilemma arises when we take for granted the rule for real numbers in the quotient operation as $\frac{\sqrt{a}}{\sqrt{b}} = \sqrt{\frac{a}{b}}$, and carelessly extend it to negative numbers.

The following is clearly true since both sides of the equation are equal to $\sqrt{-1}$. Now watch what happens when we accept the above generalization: Starting with $\sqrt{\frac{1}{-1}} = \sqrt{\frac{-1}{1}}$ would then lead to $\frac{\sqrt{1}}{\sqrt{-1}} = \frac{\sqrt{-1}}{\sqrt{1}}$.

Clearing fractions (perhaps by either multiplying by the common denominator or by simply cross-multiplying), we get $\left(\sqrt{1}\right)^2 = \left(\sqrt{-1}\right)^2$. This essentially tells us that $1 = -1$. Again, the definition was abused, leading to a mistaken result. To debunk this "proof" one need not know much about complex numbers, just a familiarity with the characteristics of the familiar operations. We notice that these are times that our time-honored operations take on other characteristics.

A SUBTLE MISTAKE LEADS TO AN ABSURDITY

Follow along as we solve this equation for x (a real number):

$$\frac{6}{x-3}-\frac{9}{x-2}=\frac{1}{x-4}-\frac{4}{x-1}.$$

We begin by multiplying the fractions on each side by their respective lowest common denominator:

$$\frac{6(x-2)}{(x-2)(x-3)}-\frac{9(x-3)}{(x-2)(x-3)}=\frac{x-1}{(x-1)(x-4)}-\frac{4(x-4)}{(x-1)(x-4)}.$$

We then clear the parentheses and add the fractions on each side of the equation:

$$\frac{6x-12-9x+27}{x^2-3x-2x+6}=\frac{x-1-4x+16}{x^2-4x-x+4}.$$

By combining like terms, we get:

$$\frac{-3x+15}{x^2-5x+6}=\frac{-3x+15}{x^2-5x+4}.$$

Now divide both sides by $(-3x+15)$:

$$\frac{1}{x^2-5x+6}=\frac{1}{x^2-5x+4}.$$

We then equate denominators, since the numerators and the fractions are equal: $x^2-5x+6=x^2-5x+4$. By subtracting x^2-5x from both sides of the equation, we end up with $6=4$.

With this absurd result, you would think that the original equation has no solution. This is wrong! The solution of this equation is $x=5$, as you can see from the following, where we show that when $x=5$, each side of the original equation has the same value, namely, 0:

$$\frac{6}{5-3}-\frac{9}{5-2}=\frac{6}{2}-\frac{9}{3}=3-3=0, \text{ and } \frac{1}{5-4}-\frac{4}{5-1}=\frac{1}{1}-\frac{4}{4}=1-1=0.$$

We notice that x cannot take on the values of 1, 2, 3, and 4, since that would produce a zero denominator in one of the fractions of the original equation.

So, then, where might the error lie? When we divided by $-3x + 15$, we had to eliminate the possibility that $-3x + 15 = 0$. However, this case is the one that provides us with the correct answer, since $3x = 15$, and, therefore, $x = 5$. Therefore, we once again—surprisingly—divided by zero. Our old nemesis!

A CONFUSING EQUATION RIPE FOR A MISTAKE (OR AN OMISSION)

We seek to solve the equation for the real number x: $3 - \frac{2}{1+x} = \frac{3x+1}{2-x}$.

Add the terms on the left side of the equation to get: $\frac{3(1+x)}{1+x} - \frac{2}{1+x} = \frac{3x+1}{1+x}$.

This gives us: $\frac{3x+1}{1+x} = \frac{3x+1}{2-x}$. Since the numerators are equal, the denominators must be equal as well. $1 + x = 2 - x$. Solving for x, we get $2x = 1$, and $x = \frac{1}{2}$. This would appear to be a solution to the equation. Let's check to see if this checks out properly. Substituting the $\frac{1}{2}$ for x, we get the following:

The left side of the equation: $3 - \frac{2}{1+\frac{1}{2}} = 3 - \frac{2}{\frac{3}{2}} = 3 - \frac{4}{3} = \frac{5}{3}$.

The right side of the equation: $\frac{3\cdot\frac{1}{2}+1}{2-\frac{1}{2}} = \frac{\frac{5}{2}}{\frac{3}{2}} = \frac{5}{3}$. All appears to be fine.

Unfortunately, that is not the only solution to this equation.

Let's begin another method for solution of this equation:

$$3 - \frac{2}{1+x} = \frac{3x+1}{2-x}.$$

Multiply both sides of the equation by $(1 + x)(2 - x)$ to get $3 \cdot (1 + x)(2 - x) - 2 \cdot (2 - x) = (3x + 1)(1 + x)$. Then, clearing parentheses, we get $-3x^2 + 3x + 6 - 4 + 2x = 3x^2 + 3x + x + 1$. Then, simplifying, $-3x^2 + 5x + 2 = 3x^2 + 4x + 1$. We will now add $3x^2 - 5x - 2$ to both sides of the equation: $0 = 6x^2 - x - 1$. By dividing both sides of the equation by 6, we then have:

$$x^2 - \frac{1}{6}x - \frac{1}{6} = 0.$$

Using the well-known quadratic formula to solve this quadratic equation, we get the two values for x.

$x_{1,2} = \frac{1}{12} \pm \sqrt{\frac{1}{12^2}+\frac{1}{6}} = \frac{1}{12} \pm \sqrt{\frac{1}{144}+\frac{24}{144}} = \frac{1}{12} \pm \frac{5}{12}$, or written separately:

$$x_1 = \frac{1}{12} + \frac{5}{12} = \frac{1}{2}, \text{ and } x_2 = \frac{1}{12} - \frac{5}{12} = -\frac{1}{3}.$$

$x_2 = -\frac{1}{3}$ is then a second solution to this equation (along with the previously found $x_1 = \frac{1}{2}$).

We ought to check the second solution to see if it is, in fact, a correct solution.

Substituting the second solution, $x_2 = -\frac{1}{3}$, on the left side of the original equation:

$$3 - \frac{2}{1-\frac{1}{3}} = 3 - \frac{2}{\frac{2}{3}} = 3 - 3 = 0.$$

Then, substituting the value $x_2 = -\frac{1}{3}$ on the right side of the equation, we have:

$$\frac{3\cdot\left(-\frac{1}{3}\right)+1}{2+\frac{1}{3}} = \frac{0}{\frac{7}{3}} = 0.$$

So clearly the second solution is also correct. Our mistake here was to solve an equation and to be satisfied with only one of two possible correct solutions.

Another analogous situation that merits attention is the solution for x in the following equation (with $a, b \in \mathbf{R}$): $\frac{a-x}{1-ax} = \frac{1-bx}{b-x}$.

Follow the steps below to see where there might be a mistake:

$$(a-x)(b-x) = (1-ax)(1-bx)$$
$$ab - ax - bx + x^2 = 1 - bx - ax + abx^2$$
$$x^2 = 1 + abx^2 - ab$$
$$x^2(1-ab) = 1 - ab.$$

If $1 - ab = 0$, then every x satisfies the equation. If $1 - ab \neq 0$, then $x^2 = 1$, and therefore $x = 1$ or $x = -1$. Both of these values are actually solutions, which we can verify by substitution:

$$\frac{a+1}{1+a} = \frac{1+b}{b+1}\text{; therefore, } a \neq -1 \text{ and } b \neq -1,$$

$$\text{just as } \frac{a-1}{1-a} = \frac{1-b}{b-1} \text{ implies } a \neq 1 \text{ and } b \neq 1.$$

But what does $1 = ab$ mean?

When a and b are in a reciprocal relationship, then x may take on any value.

When we substitute $b = \frac{1}{a}$ into the original equation,

$$\frac{a-x}{1-ax} = \frac{1-\frac{x}{a}}{\frac{1}{a}-x} = \frac{a-x}{1-ax},$$

we see that both sides are identical. This would imply that every value of x is a valid solution. Yet, when $x = \frac{1}{a}$, the denominator will equal 0.

The complete result, then, is:

Solution 1: $a \neq -1$ and $b \neq -1$: yields $x = -1$.
Solution 2: $a \neq 1$ and $b \neq 1$: yields $x = 1$.
Solution 3: $ab = 1$: all values of x, where $|x| \neq 1$ and $x \neq \frac{1}{a}$ (as well as $x \neq \frac{1}{b}$).

AN EQUATION THAT LEADS TO A MISTAKEN SOLUTION

Suppose we search for the solution (x, y) of the equation $x^2 + 9y^2 = 0$, where x and y are real numbers.

Follow the solution below:

$$x^2 + 9y^2 = 0$$

Subtract x^2 from both sides of the equation: $9y^2 = -x^2$.
Divide both sides equation by 9: $y^2 = -\frac{x^2}{9}$.
Take the square root of both sides of the equation: $y = \pm\frac{x}{3}$.
This is a mistake—not the correct solution.

Since $x^2 \geq 0$ and $9y^2 \geq 0$, we know that $x^2 + 9y^2 \geq 0$. However, we are-given that $x^2 + 9y^2 = 0$. Therefore, it must follow that $x^2 = 0$ and $9y^2 = 0$, which implies that $x = 0$, as well as $y = 0$, and not $y = \pm\frac{x}{3}$.

(The original equation has no solution in the real numbers, yet, in the complex numbers, the following is a solution: $y = \frac{i}{3}x$, or $y = -\frac{i}{3}x$.)

CAUTION MUST BE TAKEN WHEN SOLVING INEQUALITIES: OTHERWISE WE WILL MAKE A MISTAKE

We begin with the real numbers a and b. For which values of a and b is the following inequality satisfied?

$$\frac{a}{b} + \frac{b}{a} > 2$$

We realize that the values of a and b cannot be zero, or else the division indicated would be invalid.

We begin by multiplying both sides by ab to get: $a^2 + b^2 > 2ab$. Then add to both sides $-ab - b^2$ to get $a^2 - ab > ab - b^2$. Factoring the common factor on both sides of the equation gives the following: $a(a - b) > b(a - b)$.

Our last step is to divide both sides by $(a - b)$, which results in $a > b$.

The solution to this inequality appears to be $a > b$. Is this answer correct?

Clearly, with the division by $(a - b)$, we realized that $a \neq b$. If $a = b$, then following would be true:

$$\frac{a}{a} + \frac{a}{a} = 1 + 1 = 2, \text{ which contradicts the given inequality.}$$

So now let's look at an alternate solution to this inequality.

$$\frac{a}{b} + \frac{b}{a} > 2$$

We begin by multiplying both sides by ab: $a^2 + b^2 > 2ab$. Then add $-2ab$ both sides of the inequality: $a^2 - 2ab + b^2 > 0$. Factoring gives us the following: $(a - b)^2 > 0$. Taking the square root of both sides gives us an absolute value on the left side: $|a - b| > 0$. This reduces the following two possibilities:

$|a-b| = a - b > 0$, which implies that $a > b$; and
$|a-b| = -(a-b) = -a + b > 0$, which implies that $a < b$.

Once again for $a \neq b$, we have found a solution to the inequality.

The correct answer is the following:

(a) $a \neq b$ and $a, b > 0$; and
(b) $a \neq b$ and $a, b < 0$.

In other words, when a and b have different signs, the inequality is not satisfied. This can be seen graphically in figure 3.6.

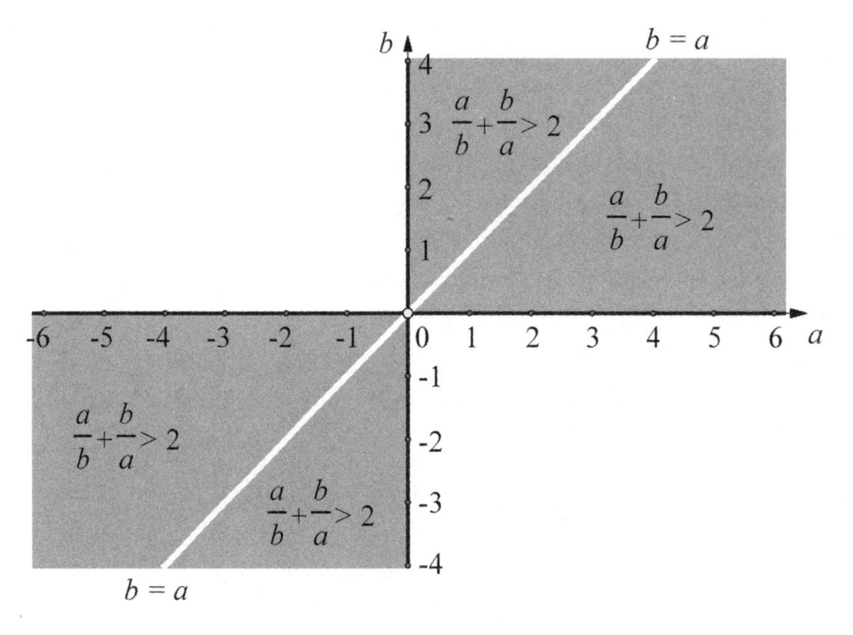

Figure 3.6.

FURTHER MISTAKES WITH INEQUALITIES

A common error with inequalities occurs when reciprocals are formed, or when they are multiplied by a negative number. In these two cases, the inequality symbol must be reversed, as is shown in the following situations:

Beginning with $2 < 3$, we have to reverse the inequality symbol for $\frac{1}{2} > \frac{1}{3}$, and for $-2 > -3$.

Here is an example of such a common mistake, where we seek to find the value of n. (Where n is a natural or real number.)

$$n \cdot \log 0.1 < \log 0.01$$

We then divide both sides by log 0.1 and get:

$$n < \frac{\log 0.01}{\log 0.1} = \frac{-2}{-1} = 2, \text{ or simply written } n < 2.$$

However, logic tells us that $n > 2$.

Recall what the value of log 0.1 is, and you will see the mistake. (Yes, it is negative!)

MISTAKES THAT LEAD TO CORRECT RESULTS

There are times when weird and illegitimate algorithms lead to correct results, as in the following illustrations. (You may wish to compare this to the example of the mistaken fraction reduction in chapter 2 ($\frac{16}{64} = \frac{1\not{6}}{\not{6}4} = \frac{1}{4}$). Notice that you cannot simply remove a number from under the radical sign and keep the value the same—yet here are some examples where this mistake leads to a correct value.

$$\sqrt{2\frac{2}{3}} = 2 \cdot \sqrt{\frac{2}{3}},$$

$$\sqrt{3\frac{3}{8}} = 3 \cdot \sqrt{\frac{3}{8}},$$

$$\sqrt{4\frac{4}{15}}=4\cdot\sqrt{\frac{4}{15}},$$

$$\sqrt{5\frac{5}{24}}=5\cdot\sqrt{\frac{5}{24}},$$

$$\sqrt{12\frac{12}{143}}=12\cdot\sqrt{\frac{12}{143}},$$

$$\sqrt[3]{2\frac{2}{7}}=2\cdot\sqrt[3]{\frac{2}{7}},$$

$$\sqrt[3]{3\frac{3}{26}}=3\cdot\sqrt[3]{\frac{3}{26}}, \text{ and so on.}$$

We might want to know under what conditions this will be true. That is, when will $\sqrt[n]{a+b}=a\sqrt[n]{b}$. If we take the nth power of both sides, we get $a + b = a^n \cdot b$. Then the following steps lead us to the relationship between a and b:

$$b - ba^n = -a$$
$$b(1 - a^n) = -a$$
$$b=\frac{a}{a^n-1}.$$

For the ambitious reader we provide some explanation of this phenomenon. The general case looks like this:

$$\sqrt[n]{a+\frac{a}{a^n-1}}=a\cdot\sqrt[n]{\frac{a}{a^n-1}}$$

and can be justified for the case where $a > 1$, as follows:

$$\sqrt[n]{a+\frac{a}{a^n-1}}=\left(a+\frac{a}{a^n-1}\right)^{\frac{1}{n}}=\left(\frac{a(a^n-1)}{a^n-1}+\frac{a}{a^n-1}\right)^{\frac{1}{n}}=\left(\frac{a(a^n-1+1)}{a^n-1}\right)^{\frac{1}{n}}$$

$$=\left(\frac{a\cdot a^n}{a^n-1}\right)^{\frac{1}{n}}=a\cdot\left(\frac{a}{a^n-1}\right)^{\frac{1}{n}}=a\cdot\sqrt[n]{\frac{a}{a^n-1}}.$$

However, caution must be used before generalizing this further—that is when mistakes occur.

While on the topic of weird radical reductions, we shall consider the following, which appears to be correct:

$$\sqrt{2^2+\frac{4}{3}} = 2 \cdot \sqrt{\frac{4}{3}},$$

$$\text{or } \sqrt{3^2+\frac{9}{8}} = 3 \cdot \sqrt{\frac{9}{8}}.$$

The question is, then, can this be generalized in this way $\sqrt{a^2+b} = a \cdot \sqrt{b}$?

Unfortunately not. Yet, we would wonder for which cases such strange square-root extraction would be possible.

Let's consider this case: $\sqrt{a^2+b} = a \cdot \sqrt{b}$. Squaring both sides, we get $a^2 + b = a^2 \cdot b$. Then, rearranging the terms, $a^2 = a^2 \cdot b - b = b(a^2 - 1)$. And then, solving for b, we find $b = \frac{a^2}{a^2-1}$, which tells us the restriction for which this can work. One might say that these are "maginificent mistakes," since they are mistakenly done, and yet lead to a correct value.

MORE MISTAKES THAT LEAD TO CORRECT RESULTS

As we have seen from the previous example, mistakes don't always lead to an absurd result; we could also have mistakes that lead to a correct answer. These are not to be condoned, but just provide us with some amusement.

We begin with the equation $x - 2 = 3$, which is the same as $x = 5$. Now we will make a deliberate mistake and add 12 to *only* the left side of the original equation to get $x + 10 = 3$. Then we will multiply both sides of the equation by $x - 5$ to get $(x + 10)(x - 5) = 3(x - 5)$. We now subtract $3(x - 5)$ from both sides of the equation, which gives us $x^2 + 5x - 50 - (3x - 15) = 0$, or in simplified form $x^2 + 2x - 35 = 0$. By factoring, we get $(x + 7)(x - 5) = 0$. Dividing both sides by $x + 7$ gives us $x - 5 = 0$, or $x = 5$, which is what we had as our initial value of x. So, despite our earlier mistake of adding 12 to only one side of the equation, we still got the right result.

Had we not added 12 to only one side of the equation, but to both sides, as we should, we would have subtracted $15(x - 5)$ instead of $3(x - 5)$. This would have given us $(x - 5)^2 = 0$, implying that $x = 5$. The "wrong" solution $x = -7$, will have disappeared through the division by $x + 7$.

Another comical mistake that leads to a correct answer is to incorrectly *add* the two binomials instead of following the indicated multiplication.

We begin with the equation that we are asked to solve for x.

$$(5 - 3x)(7 - 2x) = (11 - 6x)(3 - x)$$

Now *adding* instead of *multiplying*, as is indicated in the given equation, we get $(5 - 3x) + (7 - 2x) = (11 - 6x) + (3 - x)$. This can (correctly) be converted to: $12 - 5x = 14 - 7x$. This yields $2x = 2$, or $x = 1$, which, surprisingly, is correct!

Compare this to solving the equation $(5 - 3x)(7 - 2x) = (11 - 6x)(3 - x)$ correctly, which leads to $6x^2 - 31x + 35 = 6x^2 - 29x + 33$, which has the sole solution $x = 1$.[4] Here is a rather silly series of two mistakes that leads to the correct answer: $\sqrt{\frac{2.8}{70}} = \sqrt{0.4} = 0.2$. In other words, the second error corrects the first error.

A CORRECT START FOLLOWED BY A SERIES OF SILLY MISTAKES LEADS TO THE CORRECT SOLUTION

We are given the following equation and asked to solve for x:

$$\frac{x-7}{x+7} + \frac{x+10}{x+3} = 2.$$

By multiplying both sides of the equation by $(x + 7)(x + 3)$, as we would normally begin in solving this equation, we get the following. This is then followed by the mistaken cancellations shown below, which is then followed by some further sloppiness that brings us to a correct answer:

$$(x - 7) \cdot \cancel{(x+3)} + (x + 10) \cdot \cancel{(x+7)} = 2\cancel{(x+7)} \cdot \cancel{(x+3)}$$
$$x - 7 + x + 7 = 2$$
$$2x = 2$$
$$x = 1.$$

This mistake may appear comical, but it has been already committed by some.

RIGHT PROCEDURE WITH THE WRONG EQUATION

When we are given the equation $\frac{1}{x+1}+\frac{x}{x+2}+\frac{1}{x+3}=1$, the usual first step toward a solution calls for multiplying through by the lowest common denominator of the given fractions, $(x + 1)(x + 2)(x + 3)$. However, we will multiply each of the fractions by 1 in the form of

$\frac{(x+1)(x+2)(x+3)}{(x+1)(x+2)(x+3)}$. This gives us:

$$\frac{(x+2)(x+3)}{(x+1)(x+2)(x+3)}+\frac{x(x+1)(x+3)}{(x+1)(x+2)(x+3)}+\frac{(x+1)(x+2)}{(x+1)(x+2)(x+3)}=1.$$

By multiplying as indicated each of the numerators and denominators [Note that $(x + 1)(x + 2)(x + 3) = x^3 + 6x^2 + 11x + 6$.], and then combining like terms, we get:

$$\frac{x^3+6x^2+11x+8}{x^3+6x^2+11x+6}=1,$$

which then leads us to the absurd result that

$$x^3 + 6x^2 + 11x + 8 = x^3 + 6x^2 + 11x + 6,$$

which, put another way, leaves us with $8 = 6$.

Where did we go wrong here? We surely did not divide by zero anywhere. The mistake lies in the fact that this equation does not have a solution. The value of $\frac{1}{x+1}+\frac{x}{x+2}+\frac{1}{x+3}$ approaches 1 but never reaches it. You can see this graphically in figure 3.7.

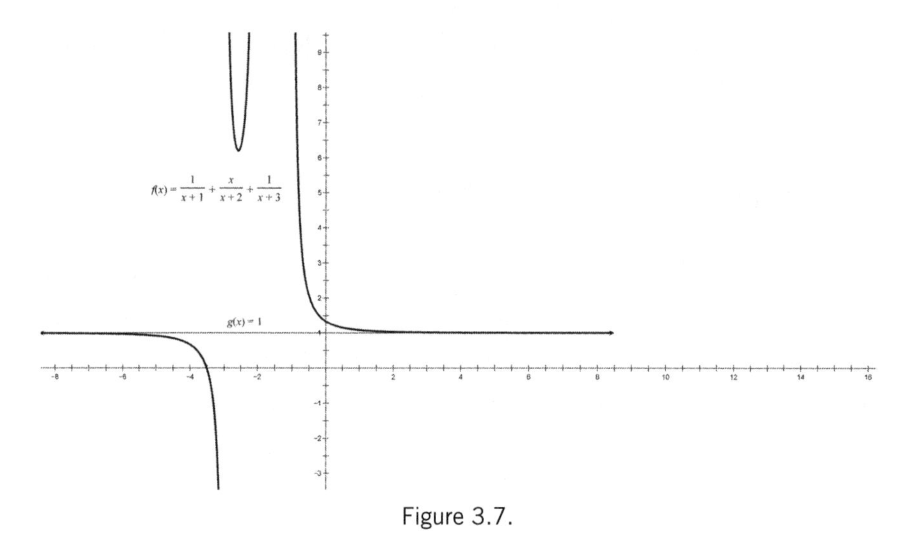

Figure 3.7.

We could also look at this situation as follows: Consider the equation

$$\frac{x}{x+1}+\frac{x}{x+2}+\frac{x}{x+3}=1 \text{ and define } y=x+2.$$

Then we get $\frac{1}{y-1}+\frac{y-2}{y}+\frac{1}{y+1}=1$ or $\frac{1}{y-1}+\frac{y}{y}-\frac{2}{y}+\frac{1}{y+1}=1$, therefore,

$$\frac{1}{y-1}-\frac{1}{y}-\frac{1}{y+1}-\frac{1}{y}=0, \text{ which gives us } \frac{1}{y(y-1)}=\frac{1}{y(y+1)}.$$

This is a contradiction. Therefore, the given equation has no solution.

A MISTAKE FOR WHICH WE CAN BLAME THE CALCULATOR

Some mistakes in mathematics may not be our fault. Rather they may be the fault of the calculator on which we seem to have unquestioned reliance. Suppose we take an algebraic fraction such as $\frac{1}{\sqrt{a+b}-\sqrt{a}}$ and correctly find its equivalent by first multiplying by 1 in the form of $\frac{\sqrt{a+b}+\sqrt{a}}{\sqrt{a+b}+\sqrt{a}}$ and then doing the algebra as shown here:

$$\frac{1}{\sqrt{a+b}-\sqrt{a}}=\frac{1}{\sqrt{a+b}-\sqrt{a}}\cdot\frac{\sqrt{a+b}+\sqrt{a}}{\sqrt{a+b}+\sqrt{a}}=\frac{\sqrt{a+b}+\sqrt{a}}{(\sqrt{a+b})^2-(\sqrt{a})^2}=\frac{\sqrt{a+b}+\sqrt{a}}{(a+b)-a}=\frac{\sqrt{a+b}+\sqrt{a}}{b}.$$

Let us now compare the way the calculator evaluates the two equal algebraic expressions:

$$\frac{1}{\sqrt{a+b}-\sqrt{a}} \text{ and } \frac{\sqrt{a+b}+\sqrt{a}}{b}.$$

Assigned Values		$\frac{1}{\sqrt{a+b}-\sqrt{a}}$	$\frac{\sqrt{a+b}+\sqrt{a}}{b}$
$a = 1{,}000$ $b = 0.001$	Calculator result to 8 places	63,2**91.139**	63,2**45.569**
	DERIVE 6 result to 20 places	63,245.5690147519**92618**	63,245.5690147519**34636**
$a = 100$ $b = 0.01$	Calculator result to 8 places	2000.**4001**	2000.**05**
	DERIVE 6 result to 20 places	2000.049998750062496**8**	2000.049998750062496**0**

Notice the differences—or we could say—notice the mistakes that the calculator has made. These are not exactly mathematical mistakes. They are rounding-off mistakes that can result in our making modern mistakes in our mathematical work.

MISTAKEN RELATIONSHIPS

An unfortunate mistake is when one misunderstands what a proportion is. The common mistake is one that can be seen when a woman is asked her age and reasons as follows: I was 20 years old when I got married to my husband, who at that time, was 30 years old. Since he is 60 years old today, which is twice 30, I suspect that I am 40 years old, which is twice 20.

In other words, the woman figured as follows: $\frac{x}{20}=\frac{60}{30}$, $x = 20 \cdot 2 = 40$.

Unfortunately, this is a mistake. This cannot be handled as a proportion, since it merely requires a constant difference. In this case, there is a difference of ten years; therefore, her the correct age is $60 - 10 = 50$.

ANOTHER ABSURDITY

In a proportion, if the first term is greater than the second term, then the third term must be greater than the fourth term. Therefore, if $ad = bc$, then $\frac{a}{b} = \frac{c}{d}$. Suppose $a > b$, then $c > d$.

If we let $a = d = 1$, and $b = c = -1$, we have satisfied the equation $ad = bc$, where $a > b$. It then should follow that $c > d$, which in this case would indicate that $-1 > 1$. This is clearly a mistake, but where is the error?

The mistake here is that we said earlier that "$\frac{a}{b} = \frac{c}{d}$. Suppose $a > b$, then $c > d$." This only holds true for positive numbers, but not in general. For example, if $\frac{a}{b} = \frac{c}{d}$, and $a > b$, we could have $\frac{5}{4} = \frac{-10}{-8}$ and $5 > 4$, but -10 is *not* greater than -8. Therefore, here $c < d$.

MISUNDERSTANDINGS WHEN ADDING EQUATIONS LEADS TO MISTAKES

We know that if we add equations, the result is also an equation.

Consider the following:

One cat has four legs (1)
Zero cats have three legs (2)

If we now had these two "equations," we should be able to conclude that one cat has seven legs, since we added the cats $(1 + 0 = 1)$ and the legs $(4 + 3 = 7)$. Clearly this is a mistake, which we see from the absurdity in the result. The error lies in the misinterpretation of the words *have* and *has* as being the same as *equals*. In fact, the above two statements are not equations and therefore cannot be treated as such.

MISUNDERSTANDING SYSTEMS OF EQUATIONS LEADS TO MISTAKES

Suppose we are given the following system of equations:

$$-x + 2y + z = -2 \quad (1)$$
$$x - y - z = 1 \quad (2)$$
$$x - 2y + z = -2 \quad (3)$$
$$x - y + z = 1 \quad (4)$$

One way to solve these equations is to add the first two and then to subtract the fourth from the second, as shown below:

$$(1) + (2): y = -1 \quad (5)$$
$$(2) - (4): z = 0 \quad (6)$$

We now substitute the values found in equations (5) and (6) into equation (3) to get:

$$x = -4. \quad (7)$$

Therefore, the solution is $x = -4$, $y = -1$, and $z = 0$.

The same system of equations can be solved in another way, as follows:

Equate equations (1) and (3), since the left sides are both equal to the same number (–2), to get:

$-x + 2y + z = x - 2y + z$, it then follows that $4y = 2x$, or $x = 2y$. (8)

Now, substituting equation (8) into equation (2): $2y - y - z = 1$, which simplifies to: $y - z = 1$. (9)

We now substitute equation (8) into equation (4):
$2y - y + z = 1$, then $y + z = 1$. (10)

By adding equations (9) and (10), we have: $2y = 2$, or $y = 1$. (11)

Subtracting equations (9) and (10), we get: $-2z = 0$, thus $z = 0$. (12)

Substituting the results of equations (11) and (12)
into equation (1): $2 - x = -2$, we have $x = 4$. (13)

This gives us the solution $x = 4$, $y = 1$, and $z = 0$, which differs from that obtained earlier.

Where, then, is the error? It turns out that both solutions are mistaken. There is no solution to this system of equations. Actually, the equations contradict each other in that they deliver conflicting values of the variables, such as $z = 0$ and $z = -2$.

A COMMON MISTAKE OF MISUNDERSTANDING GIVEN INFORMATION

Consider purchasing two pens, a red one and a typical black one. If the red one costs a dollar more than the black one, and together they cost \$1.10, what is the price of each of the two pens? Most people will (mistakenly) respond with the answer of \$1.00 for the red pen and 10¢ for the black pen (or the reverse).

To see where our mistake is in this reasoning, we represent the situation algebraically as follows: $x + (x + 100) = 110$, where x is the price of the black pen. Solving this equation, we find that $x = 5$.

Therefore, the red pen costs \$1.05, and the black pen costs \$0.05.

Jumping to conclusions without checking the reasonableness of an answer leaves one open to making a mistake.

This sort of mistake can be seen from the following example.

Consider a 100 kg barrel of a mixture where there is 99 percent water. A short time later, some of the water evaporates, reducing the water in the barrel to 98 percent. How much does the barrel weigh with this water reduction?

The typical answer—a mistaken one, at that—is that 99 percent − 98 percent = 1 percent of the water evaporates, and that 1 percent of 100 kg is 1 kg. Therefore, the barrel now weighs 99 kg. Unfortunately, this is incorrect. Where is the mistake?

Considering the following chart:

	Water Content in %	Water Content in kg	The Evaporated Water in %	The Evaporated Water in kg
First Case	99	99	1	1 (a constant!)
Second Case	98	???	2	1

From the chart, we see that 1 kg of the evaporated water represents, therefore, 2 percent, so that the barrel (100 percent) must weigh 50 kg and the remaining water must be 49 kg. This correct answer usually causes discomfort, since it is counterintuitive.

IS IT A MISTAKE TO GIVE ONLY ONE OF TWO POSSIBLE ANSWERS?

Consider the sequence of numbers 1, 2, 4, 7. What numbers must follow to keep this sequence consistent? Such questions usually come up on intelligence tests, and as you will see, a correct response could be considered mistaken when the questioner is expecting a different one.

One possible continuation of the sequence is:

1, 2, 4, 7, 11, 16, 22, 29, 37, 46, 56, 67, 79, 92, 106, . . . ,

where each time the difference between consecutive numbers increases by 1.

We can generalize this for a_n with $a_n = \frac{n(n+1)}{2} + 1$, with $n = 0, 1, 2, 3, \ldots$, where n is a natural number. This sequence also represents the maximum number of pieces formed when slicing a pancake with n cuts.

A complication manifests itself here in that there is also another interpretation for the continuation of this sequence. The next number after the given four numbers is still 11, but that's where the similarity ends. As you can see from this sequence, we now present as a possible continuation of the given four numbers,

1, 2, 4, 7, 11, 13, 14, 16, 22, 23, 26, 28, 29, 37, 44, . . .

This time you can observe that the sequence arises from the numbers n when $16n + 15$ is a prime number, as you can see from the chart below where those values of n are in bold print.

n	$16n + 15$	Prime Number		n	$16n + 15$	Prime Number
0	15	–				
1	31	yes		**16**	271	yes
2	47	yes		17	287	–
3	63	–		18	303	–
4	79	yes		19	319	–
5	95	–		20	335	–
6	111	–		21	351	–
7	127	yes		**22**	367	yes
8	143	–		**23**	383	yes
9	159	–		24	399	–
10	175	–		25	415	–
11	191	yes		**26**	431	yes
12	207	–		27	447	–
13	223	yes		**28**	463	yes
14	239	yes		**29**	479	yes
15	255	–		30	495	–

We will admit that this solution is not one that you would easily find, as it is a bit convoluted, but it is correct!

Perhaps it is a mistake for psychometricians to include such series on intelligence tests, when there are other possible *correct* responses. There are even more correct responses that the ambitious reader may wish to discover.

MAKING MISTAKEN GENERALIZATIONS

Admire the equalities that we find for powers of 1, 2, 3, 4, 5, 6, and 7, which are shown below.

$$1^0 + 13^0 + 28^0 + 70^0 + 82^0 + 124^0 + 139^0 + 151^0 = 4^0 + 7^0 + 34^0 + 61^0 + 91^0 + 118^0 + 145^0 + 148^0$$
$$1^1 + 13^1 + 28^1 + 70^1 + 82^1 + 124^1 + 139^1 + 151^1 = 4^1 + 7^1 + 34^1 + 61^1 + 91^1 + 118^1 + 145^1 + 148^1$$
$$1^2 + 13^2 + 28^2 + 70^2 + 82^2 + 124^2 + 139^2 + 151^2 = 4^2 + 7^2 + 34^2 + 61^2 + 91^2 + 118^2 + 145^2 + 148^2$$
$$1^3 + 13^3 + 28^3 + 70^3 + 82^3 + 124^3 + 139^3 + 151^3 = 4^3 + 7^3 + 34^3 + 61^3 + 91^3 + 118^3 + 145^3 + 148^3$$
$$1^4 + 13^4 + 28^4 + 70^4 + 82^4 + 124^4 + 139^4 + 151^4 = 4^4 + 7^4 + 34^4 + 61^4 + 91^4 + 118^4 + 145^4 + 148^4$$
$$1^5 + 13^5 + 28^5 + 70^5 + 82^5 + 124^5 + 139^5 + 151^5 = 4^5 + 7^5 + 34^5 + 61^5 + 91^5 + 118^5 + 145^5 + 148^5$$
$$1^6 + 13^6 + 28^6 + 70^6 + 82^6 + 124^6 + 139^6 + 151^6 = 4^6 + 7^6 + 34^6 + 61^6 + 91^6 + 118^6 + 145^6 + 148^6$$
$$1^7 + 13^7 + 28^7 + 70^7 + 82^7 + 124^7 + 139^7 + 151^7 = 4^7 + 7^7 + 34^7 + 61^7 + 91^7 + 118^7 + 145^7 + 148^7$$

From the seven examples, one could easily form the following conclusion, namely that for the natural number n, the following should hold:

$$1^n + 13^n + 28^n + 70^n + 82^n + 124^n + 139^n + 151^n = 4^n + 7^n + 34^n + 61^n + 91^n + 118^n + 145^n + 148^n.$$

These values are shown in the table below:

n	Sums
0	8
1	608
2	70,076
3	8,953,712
4	1,199,473,412
5	165,113,501,168
6	23,123,818,467,476
7	3,276,429,220,606,352

To make this generalization would be a predictable behavior. However, at the same time, it would also be a marvelous mistake. This mistake does not manifest itself until we take the next case, namely, where $n = 8$.

Notice the two sums that we get are no longer the same:

$$1^8 + 13^8 + 28^8 + 70^8 + 82^8 + 124^8 + 139^8 + 151^8 = 468{,}150{,}771{,}944{,}932{,}292.$$

However, $4^8 + 7^8 + 34^8 + 61^8 + 91^8 + 118^8 + 145^8 + 148^8 = 468{,}087{,}218{,}970{,}647{,}492$. As a matter of fact, the difference between these two sums is:

$$468{,}150{,}771{,}944{,}932{,}292 - 468{,}087{,}218{,}970{,}647{,}492 = 63{,}552{,}974{,}284{,}800.$$

As n increases, so does the difference between the two sums. For $n = 20$, the difference is 3,388,331,687,715,737,094,794,416,650,060,343,026,048,000. Therefore, to avoid such mistakes, one must be sure to prove a generalization before accepting it inductively.

A COMMON MISTAKE WITH MATHEMATICAL INDUCTION

A good algebra course in high school introduces students to the process of mathematical induction. This is typically used to see if a relationship is true for all cases. One begins by showing the relationship true for the case 1, 2, and so forth. Then, assuming the relationship holds true for the kth term, one must show that is also true for the (k+1)st term, thus demonstrating that it must hold true for all terms from the first onward.

To demonstrate a common mistake using mathematical induction, we consider the following:

We would like to show that for all values of n, the following is true: $2^n > 2n + 1$.

So if we accept that for $n = k$, a natural number: $2^k > 2k + 1$ is true, then we must show that this statement is also true for $n = k + 1$.

We know that for the natural number k ($k > 0$) it is true that $2^k \geq 2$.

We now add this to our original equation to get $2^k + 2^k > 2k + 1 + 2$, which can be rewritten as: $2^k \cdot 2 = 2^{(k+1)} > 2(k + 1) + 1$, which, according to mathematical induction, proves the relationship.

However, there must be a mistake somewhere, since for $n = 0$, $n = 1$, and $n = 2$, the statement is false. It only becomes true when $n = 3$ and greater.[5]

This mistake occurred when we began with $n = k$, rather than 0, 1, and 2, something that can be seen as an "oversight"!

A MISTAKE BASED ON PREMATURELY JUMPING TO CONCLUSIONS

Suppose you take a circle, put some dots along the outside, and then connect them. If only two lines cross at any point, into how many regions will the circle be divided?[6]

As we close this chapter on arithmetic mistakes, we should notice that sometimes what is seen as a mistake may, in fact, not be a mistake at all. Consider the following sequence and ask that for the next number: **1, 2, 4, 8, 16**. Most people would assume that 32 will be the next number. Yes, that would be fine. However, when the next number is given as 31 (instead of the expected 32), cries of "wrong!" are usually heard.

Much to your amazement, this is can also be a correct answer, and **1, 2, 4, 8, 16, 31** can be a legitimate sequence, and *not* a mistake!

The task now is to be convinced of the legitimacy of this sequence. It would be nice if it could be done geometrically, as that would give evidence of a physical nature. We will do that later, but in the meantime let us first find the succeeding numbers in this "curious" sequence.

We shall set up a table of differences (i.e., a chart showing the differences between terms of a sequence), beginning with the given sequence up to 31, and then work backward once a pattern is established (here to the third differences).

Original Sequence	**1**		**2**		**4**		**8**		**16**		**31**
First Difference		1		2		4		8		15	
Second Difference			1		2		4		7		
Third Difference				1		2		3			
Fourth Difference					**1**		**1**				

With the fourth differences forming a sequence of constants, we can reverse the process (turn the table upside down), and extend the third differences a few more steps with 4 and 5.

Fourth Difference					1		1		**1**		**1**				
Third Difference				1		2		3		**4**		**5**			
Second Difference			1		2		4		7		**11**		**16**		
First Difference		1		2		4		8		15		**26**		**42**	
Original Sequence	1		2		4		8		16		31		**57**		**99**

The bold-type numbers are those that were obtained by working backward from the third-difference sequence. Thus, the next numbers of the given sequence are 57 and 99. The general term is a fourth-power expression since we had to go to the third differences to get a constant.

The general term (for a given natural number n) is:

$$\frac{n^4 - 6n^3 + 23n^2 - 18n + 24}{24}.$$

To be further convinced that this sequence is legitimate and not resulting from mistakenly replacing the "32" with a "31," we shall consider the Pascal triangle. This triangle is formed by beginning on top with 1, then the second row has 1, 1, and the third row is obtained by placing 1s at the end and adding the two numbers in the second row (1 + 1 = 2) to get the 2; the fourth row is obtained the same way. After the end 1s are placed, the 3s are gotten from the sum of the two numbers above (to the right and left), that is, 1 + 2 = 3, and 2 + 1 = 3.

1
1 1
1 2 1
1 3 3 1
1 4 6 4 1
1\5 10 10 5 1
1 6 \15 20 15 6 1
1 7 21 \35 35 21 7 1
1 8 28 56\70 56 28 8 1

The horizontal sums of the rows of the Pascal triangle to the right of the bold line drawn: 1, 2, 4, 8, 16, 31, 57, 99, 163. This is again our newly developed sequence.

A geometric interpretation should further support the legitimacy of this sequence and support the beauty and consistency inherent in mathematics. To do this, we shall make a chart (see figure 3.8) of the number of regions into which a circle can be partitioned by joining points on the circle, where no three lines meet at one point; otherwise a region would be lost.

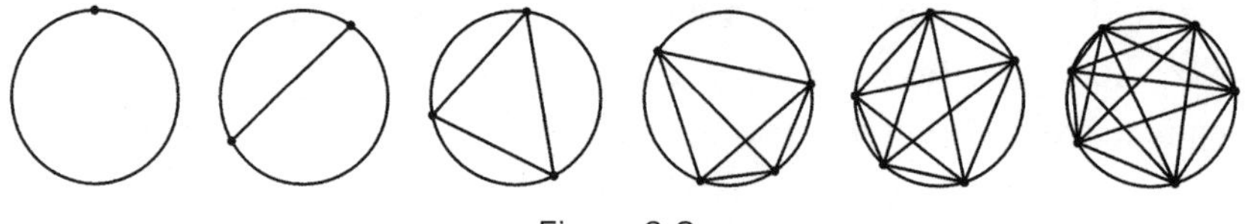

Figure 3.8.

Let's focus on the case where $n = 6$. (See figure 3.9.)

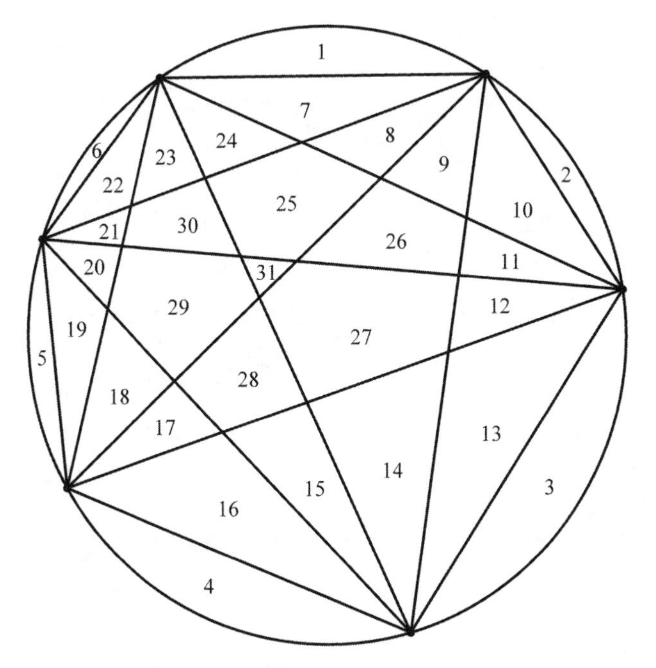

Figure 3.9.

We notice there is no thirty-second region.

Number of Points on the Circle	Number of Regions into Which the Circle Is Partitioned
1	1
2	2
3	4
4	8
5	16
6	31
7	57
8	99

Now that you see that this unusual sequence (1, 2, 4, 8, 16, 31, 57, 99, 163 . . .) appears in various other contexts, you should be convinced that even though there appeared to be a mistake at the original introduction of the "31," there was, in fact, no mistake. Thus, even mistakes can be deceptive—or mistakenly identified as mistakes!

CHAPTER 4

GEOMETRIC MISTAKES

Geometric figures can be deceiving in their depiction. We can witness this in a number of ways. For example, we can make mistakes with optical deception. Geometry is often referred to as the visual part of mathematics. We tend to believe many things as we see them. Yet geometric diagrams still play an important role in determining geometric properties and planning to prove geometric relationships. The importance of geometric diagrams should not be minimized; however, they should be carefully analyzed, as we will see throughout this chapter. Although geometric proofs can be done without seeing a diagram, picturing the geometric figures can be very helpful—but it can also be deceiving.

We can make mistakes with our visual assessment of a geometric figure. We present some of these optical tricks as they can be useful to make a person more discriminating with visual presentations. First we will show some of these easily mistaken assessments, and then we will show how logical mistakes can be compounded. So follow along as we explore some of these counterintuitive characteristics, which can lead to some magnificent mistakes!

OPTICAL MISTAKES

We can begin by comparing the two segments in figure 4.1 The one on the right side looks longer. In figure 4.2, the bottom segment also looks longer. In actuality, the segments have the same length.[1]

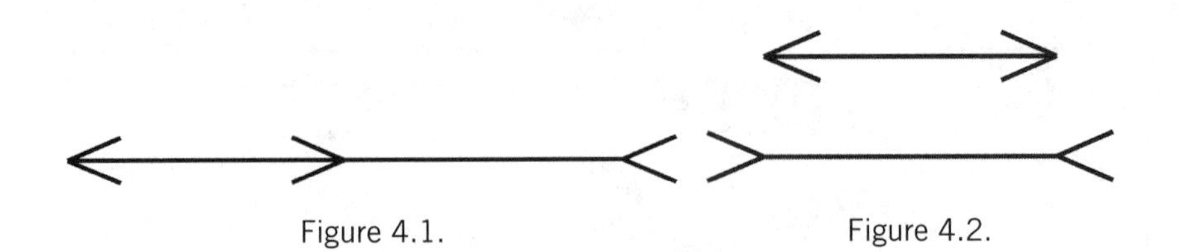

Figure 4.1. Figure 4.2.

In figure 4.3, the crosshatched segment appears longer than the clear one; and in figure 4.4, at the right side figure, the narrower and vertical stick appears to be longer than the other two, even though to the left they are shown to be the same length.

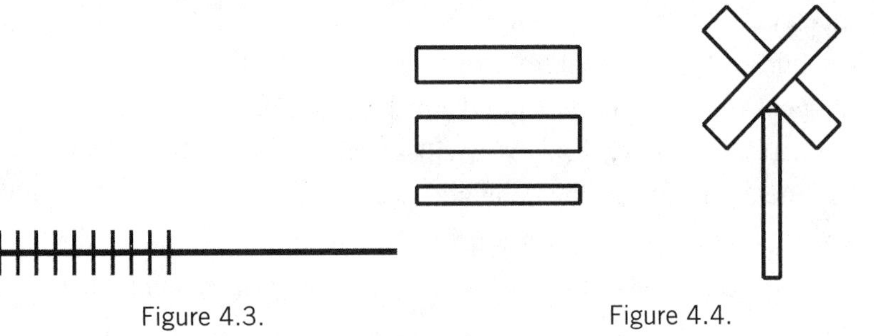

Figure 4.3. Figure 4.4.

A further optical illusion can be seen in figure 4.5, where AB appears to be longer than BC. This isn't true, since $AB = BC$.

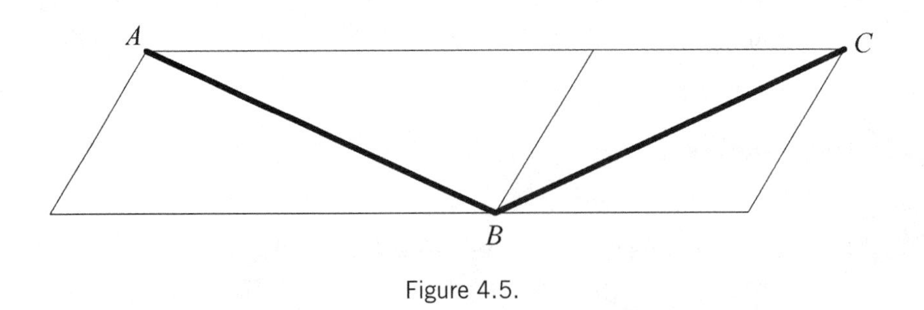

Figure 4.5.

In figure 4.6, the vertical segment clearly appears longer, but it isn't. The curve lengths and curvature of the diagrams in figure 4.7 are quite deceiving. Yet the curves are congruent!

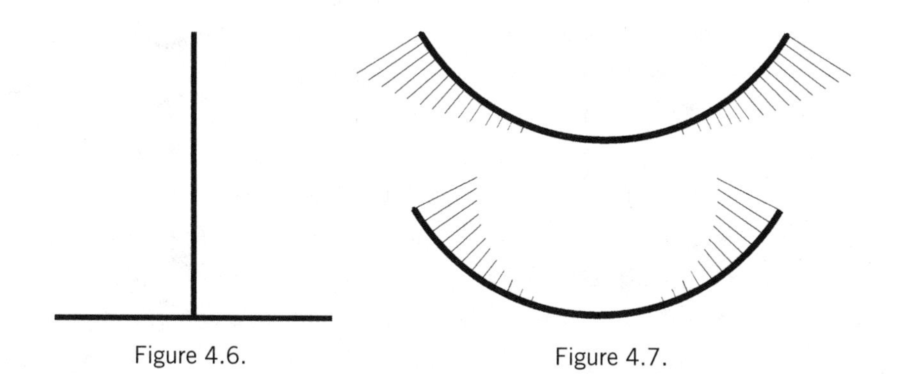

Figure 4.6. Figure 4.7.

The square between the two semicircles in figure 4.8 looks bigger than the square to the left, but the two squares are the same size. In figure 4.9, the square within the large black square looks smaller than that to the right, but, again, that is an optical illusion, since they are the same size.

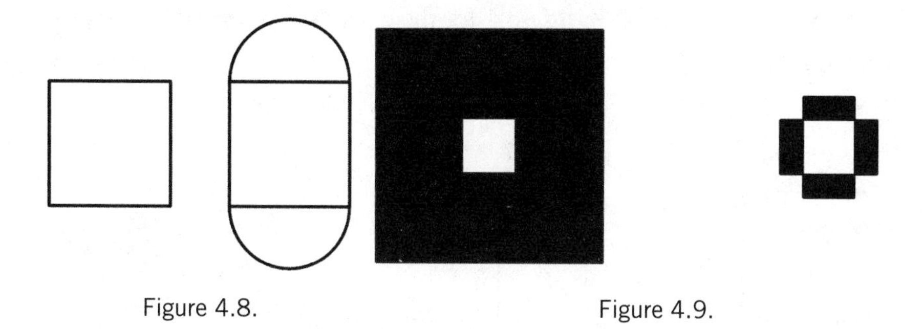

Figure 4.8. Figure 4.9.

We see further evidence of fooling the senses in figure 4.10, where the larger circle inscribed in the square (on the left) appears to be smaller than the circle circumscribed about the square (on the right). Again, the circles are the same size!

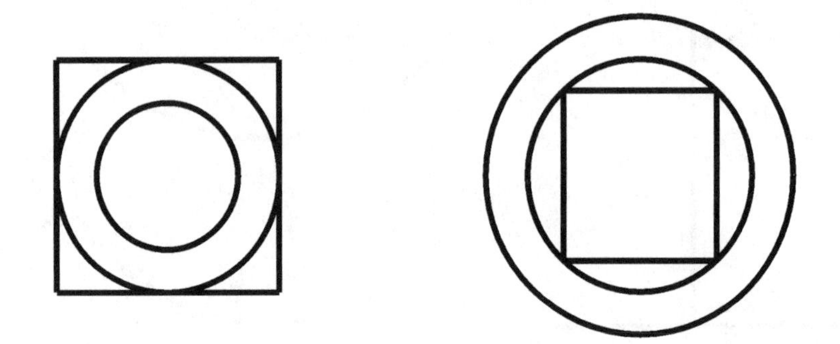

Figure 4.10.

Figures 4.11, 4.12, and 4.13 show how relative placement can affect the appearance of a geometric diagram. In figure 4.11, the center square appears to be the largest of the group, but it isn't. In figure 4.12, the black center circle on the left appears to be smaller than the black center circle on the right, and again it is not. In figure 4.13, the center sector on the left appears to be smaller than the center sector on the right. In all of these cases, the two figures that appear not to be the same size are, in fact, the same size!

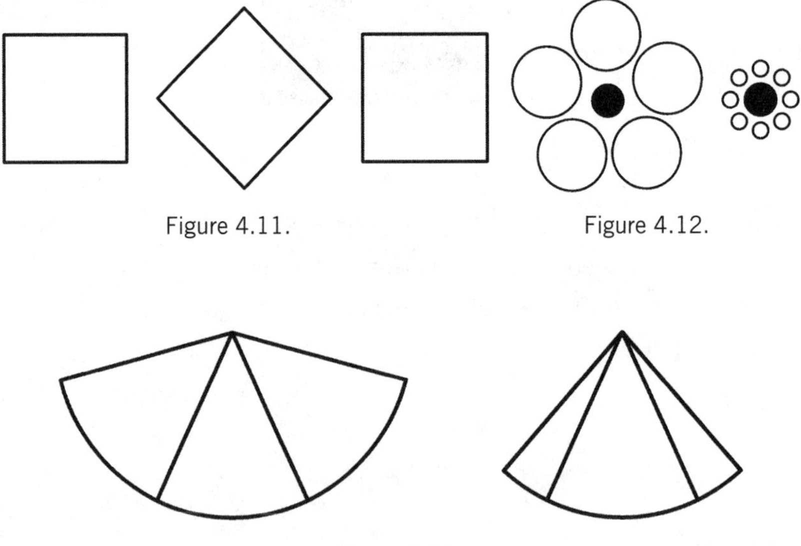

Figure 4.11.

Figure 4.12.

Figure 4.13.

In figure 4.14, the three-quarter circles are so placed as to give the visual impression that a rectangle is shown, when, in fact, there is no rectangle.

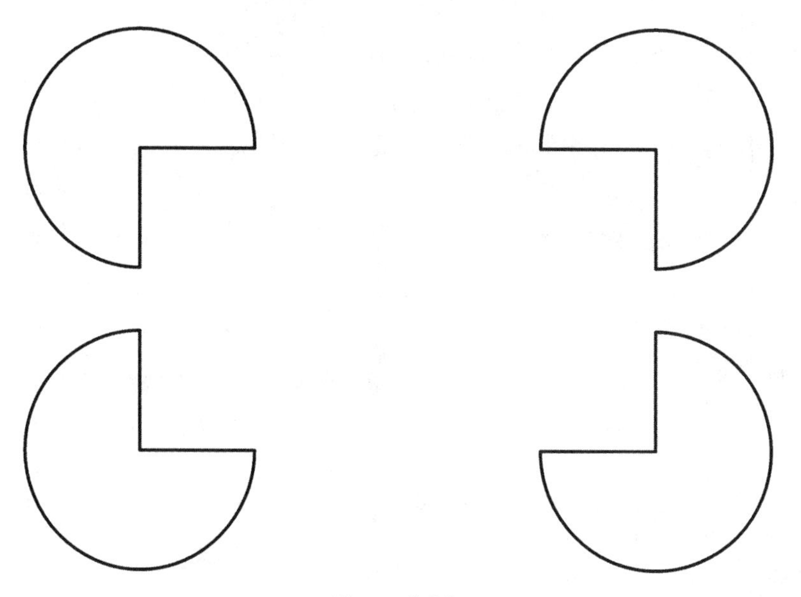

Figure 4.14.

Some optical illusions are done intentionally. Take, for example, the Penrose triangle (see figure 4.15). It appears to be a triangle with three right angles. This triangle—popularized by the English mathematician Roger Penrose (1931–) in 1958—was previously created by Oscar Reutersvärd (1915–2002), a Swedish artist, in 1934. His invention was honored by the issuance of a postage stamp in 1982 (see figure 4.16).[2]

Figure 4.15. Figure 4.16.

A postage stamp depicting an analogous figure was issued by the Republic of Austria in 1981 to commemorate the 10th International Mathematics Congress in Innsbruck (see figure 4.17).

Figure 4.17.

Just as there are optical illusions throughout our geometric world, so, too, there are "proofs" that can be fallacious; not by their errors in reasoning, but rather in their assumptions concerning geometric appearances.

MISTAKES WITH POLYGONS

It is not uncommon to try to determine the sum of the interior angles of a convex polygon by the following procedure: partition the polygon into triangles, then count the number of triangles and multiply that number by 180 to get the interior-angle sum. Take, for example, a decagon. As shown in figure 4.18, it is partitioned into eight triangles. That gives us an interior-angle sum of $8 \cdot 180° = 1{,}440°$.

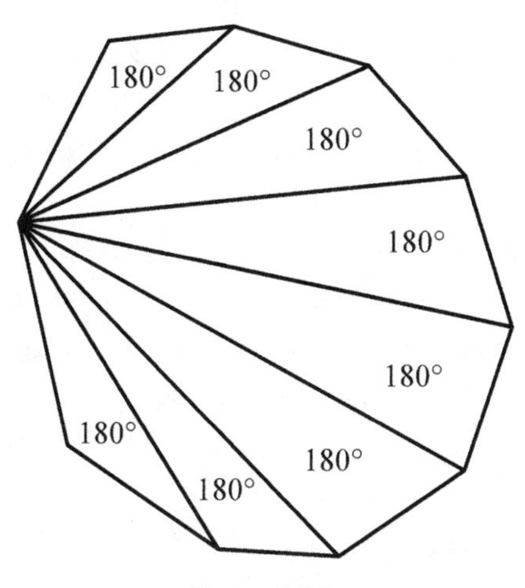

Figure 4.18.

We can also partition the decagon as shown in figure 4.19, where we draw diagonals that do not intersect each other. Again we have a partition consisting of eight triangles, which yields the correct interior-angle sum of 1,440°.

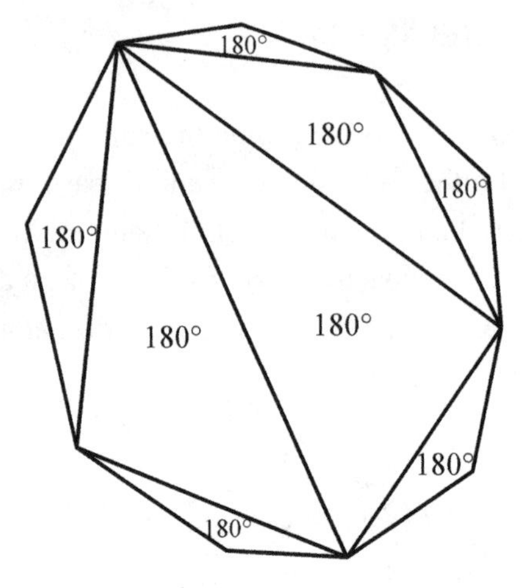

Figure 4.19.

Another attempt at partitioning the decagon for the purpose of determining the interior-angle sum can be seen in figure 4.20. Here we find that there are ten triangles to consider. This then suggests that the angles sum as $10 \cdot 180° = 1{,}800°$, which is incorrect!

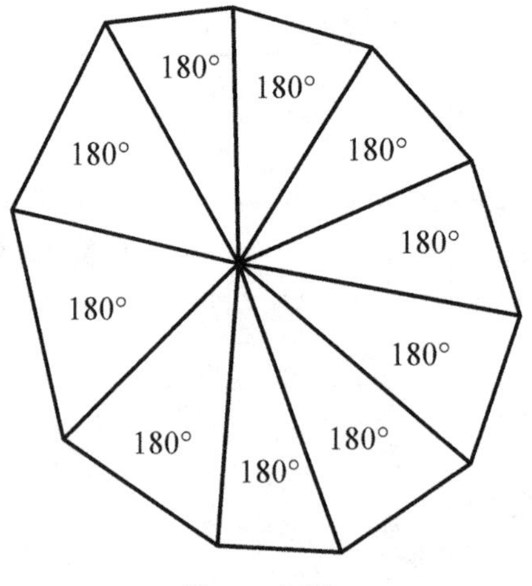

Figure 4.20.

We must now determine what the actual angle sum is. The mistake occurred in the third attempt to triangulate the decagon, where we needed to deduct the sum of the angles that are not part of the interior angles of the decagon. They are shown in figure 4.21, where we see that they have a sum of 360°, as indicated by the circle with center at their point of intersection. This shows us that it is necessary to deduct 360° from the incorrect sum of 1,800° to get the correct interior angle sum of 1,440°.

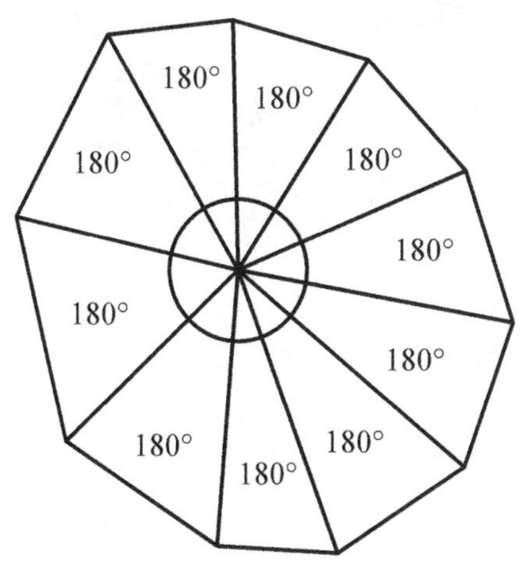

Figure 4.21.

CONFOUNDING POLYGON CONSTRUCTIONS

We are going to take a slightly different tack now. Several different constructions for a regular octagon will be presented. They will all look correct. However, we will leave it to the reader to determine which of these produces a truly regular polygon, and which are mistaken constructions—despite the fact that they look correct.

Octagon Construction 1. In figure 4.22, we begin with a square and then identify the midpoints of each of the sides of the square. Next, we join these four midpoints. At each of the four vertices of the square, an isosceles right triangle is formed. Once we bisect each of the acute angles of these isosceles right triangles, we will have identified the remaining four vertices of the octagon.

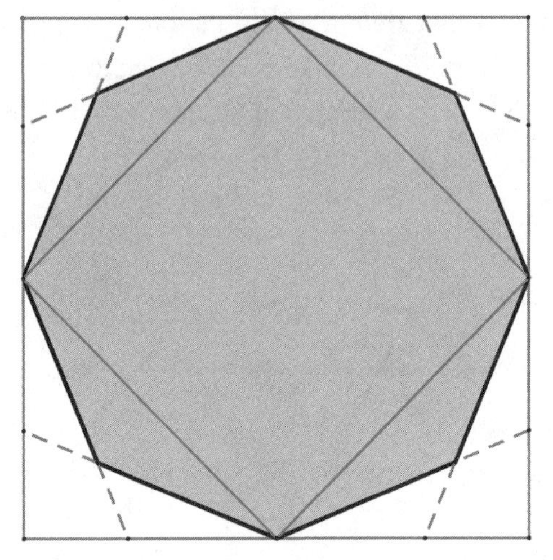

Figure 4.22.

Octagon Construction 2. Again we begin with a square and then join the midpoints of each of the sides with the square's opposite vertices (see figure 4.23).

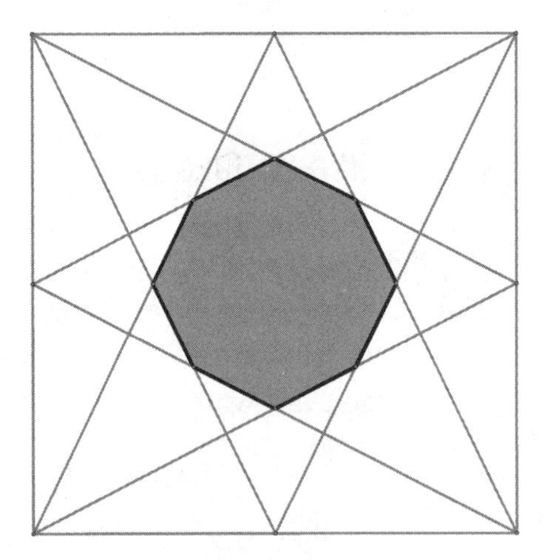

Figure 4.23.

Octagon Construction 3. We begin with four congruent tangent circles inscribed in a square as shown in figure 4.24. Next, we join the centers of each of the circles with the vertices of the square. This determines the octagon.

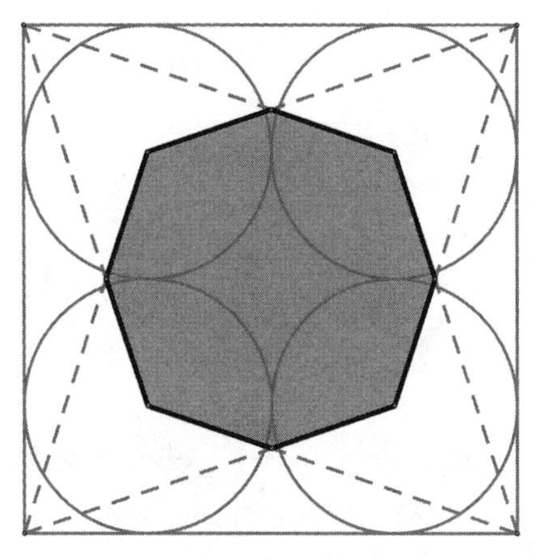

Figure 4.24.

Octagon Construction 4. Again we begin with a square. Quarter circles are drawn with center at each vertex of the square and a radius of length half the diagonal. The points at which these quarter circles intersect the sides of the square determine the required octagon (see figure 4.25).

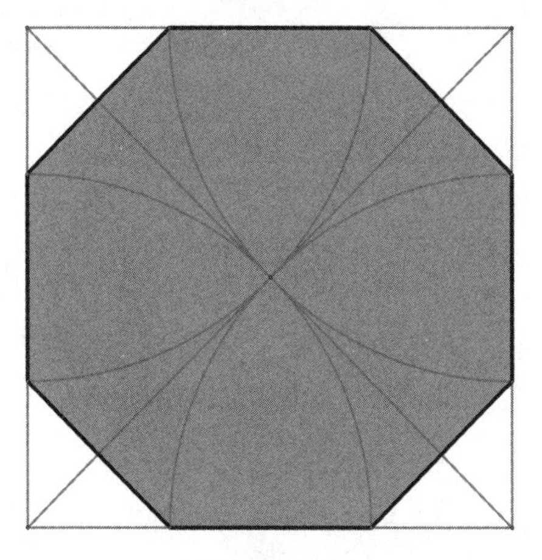

Figure 4.25.

Octagon Construction 5. Once again, begin with a square and construct a quarter circle centered at each vertex with radius the length of the side of the square. Mark the points at which the diagonals of the square intersect the quarter circles. Through these four points, lines are drawn parallel to the sides of the square, which determine the vertices of the requisite octagon (see figure 4.26).

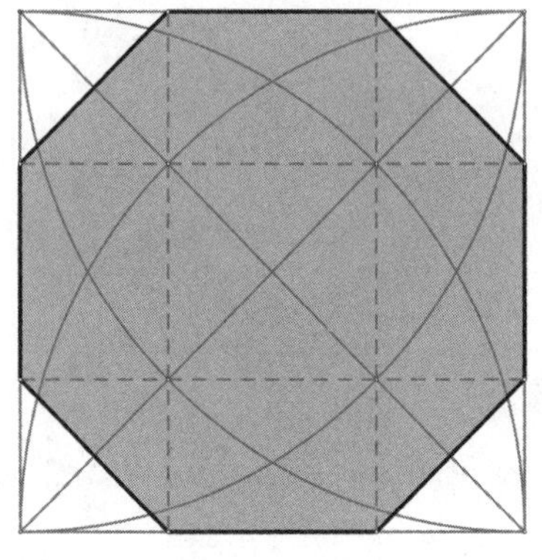

Figure 4.26.

We now have five different constructions of an octagon. The question that remains is, which of these is a regular octagon, and which of these is a mistaken construction of a regular octagon? Here are the results:

Construction 1: This is a correct construction of a regular octagon.

Construction 2: This construction was known to Archimedes (ca. 287–212 BCE) and has equal sides, but not all angles are congruent. Therefore, it is *not* a regular octagon!

Construction 3: Once again, this construction produced an octagon whose sides are of the same length, but whose angles are not of equal measure. Therefore, it is *not* a regular octagon.

Construction 4: This construction produced a regular octagon and was first developed in 1543 by the artist and geometer Augustin Hirschvogel (1503–1553).

Construction 5: This construction produced a regular octagon and was first developed in 1564 by the goldsmith Heinrich Lautensack (1522–1568).

Therefore, constructions 2 and 3 are mistaken constructions of regular octagons.

For the ambitious reader, we provide the details of each of these constructions. In the chart shown in figure 4.27, we denote the interior angles of each of the figures with the symbols φ and ψ; the side length of the octagon with b; and the side-length of the original square as a. A_{Sq} represents the area of the square.

Comparisons of the Five Octagons

(1)	(2) Archimedes	(3)	(4) Hirschvogel	(5) Lautensack
regular	Equilateral, but not equiangular	Equilateral, but not equiangular	*regular*	*regular*
$\varphi = \psi = 135°$	$\varphi \approx 126.9°$, $\psi \approx 143.1°$	$\varphi \approx 126.9°$, $\psi \approx 143.1°$	$\varphi = \psi = 135°$	$\varphi = \psi = 135°$
$b = \frac{a\sqrt{2-\sqrt{2}}}{2}$ $\approx 0.3827 \cdot a$	$b = \frac{a\sqrt{5}}{12}$ $\approx 0.1863 \cdot a$	$b = \frac{a\sqrt{10}}{12}$ $\approx 0.2635 \cdot a$	$b = a(\sqrt{2} - 1)$ $\approx 0.4142 \cdot a$	$b = a(\sqrt{2} - 1)$ $\approx 0.4142 \cdot a$
$A = \frac{a^2}{2}\sqrt{2}$ $\approx 0.7071 \cdot A_{Sq}$	$A = \frac{a^2}{6}$ $\approx 0.1667 \cdot A_{Sq}$	$A = \frac{a^2}{3}$ $\approx 0.3333 \cdot A_{Sq}$	$A = 2a^2(\sqrt{2} - 1)$ $\approx 0.8284 \cdot A_{Sq}$	$A = 2(\sqrt{2} - 1)a^2$ $\approx 0.8284 \cdot A_{Sq}$

Figure 4.27.

THE DIAGONALS OF THE HEXAGON: A MISTAKEN COUNT OF INTERSECTIONS

A common mistake when determining the number of intersections of the diagonals of a (convex) hexagon is to assume that the hexagon being considered is a regular hexagon; that is, a hexagon where all the angles and all the sides are of equal measure. We are interested in the number of intersection points of the diagonals of *any* hexagon. In figure 4.28, we can count the points of intersection of the regular hexagon shown. There are thirteen such points.

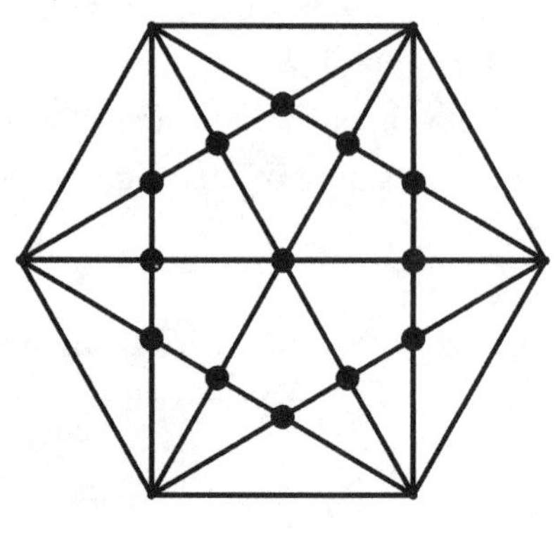

Figure 4.28.

However, if we consider an irregular hexagon, as shown in figure 4.29, we will find that the diagonals have two additional points of intersection. Thus, for a general hexagon, the number of points of intersection of the diagonals is fifteen. The mistake of using a regular hexagon when one isn't called for leads to a wrong answer.

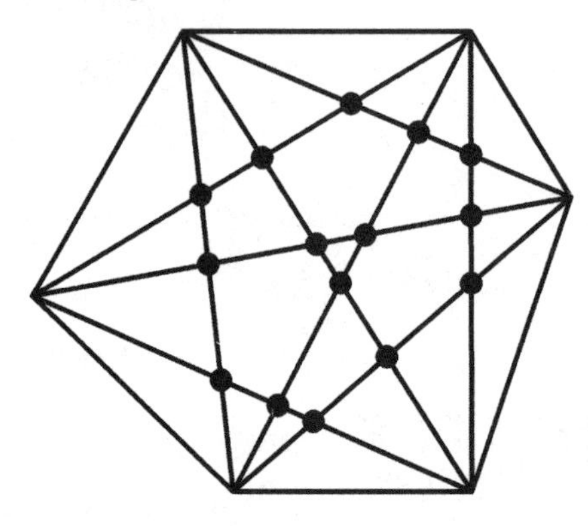

Figure 4.29.

MISTAKENLY COUNTING TRIANGLES IN A REGULAR PENTAGON

The task of counting triangles in a regular pentagon can lead to a mistaken number. This has to do with identifying different kinds of triangles and their various positions. You might want to try counting the number of triangles in the regular pentagon provided in figure 4.30. The mistake that most people make in counting, results from not setting up a systematic procedure. In this example, there are triangles that have the same shape and cause mistaken counting.

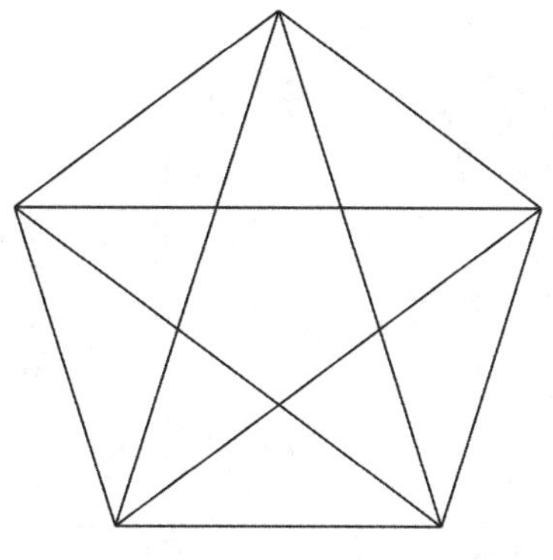

Figure 4.30.

There are, in fact, thirty-five distinct triangles that can be found by drawing the diagonals of a regular pentagon. Figure 4.31 shows the various triangles that are to be found and then counted in the figure provided.

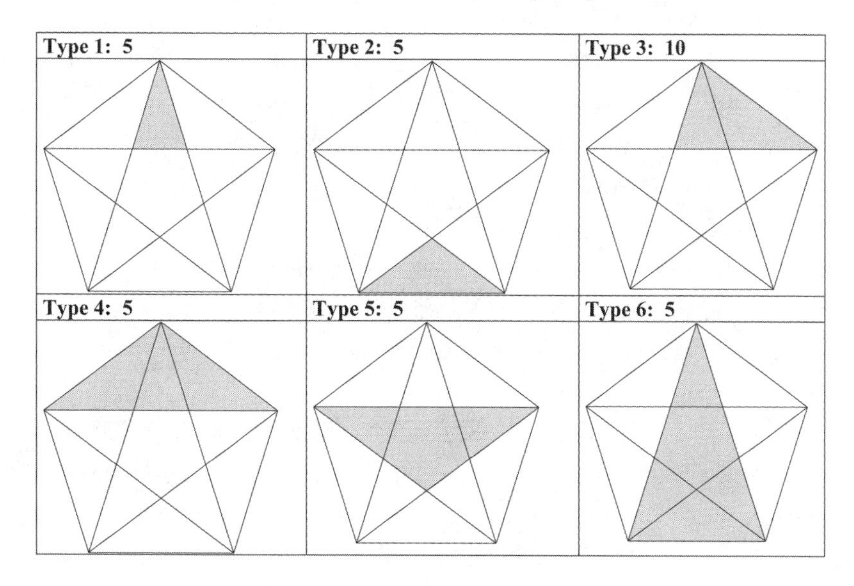

Figure 4.31.

The complete list of the thirty-five triangles is shown with their specific placements in figure 4.32. Of particular note is the systematic fashion in which the counting is done—that should avoid the commonly mistaken counting.

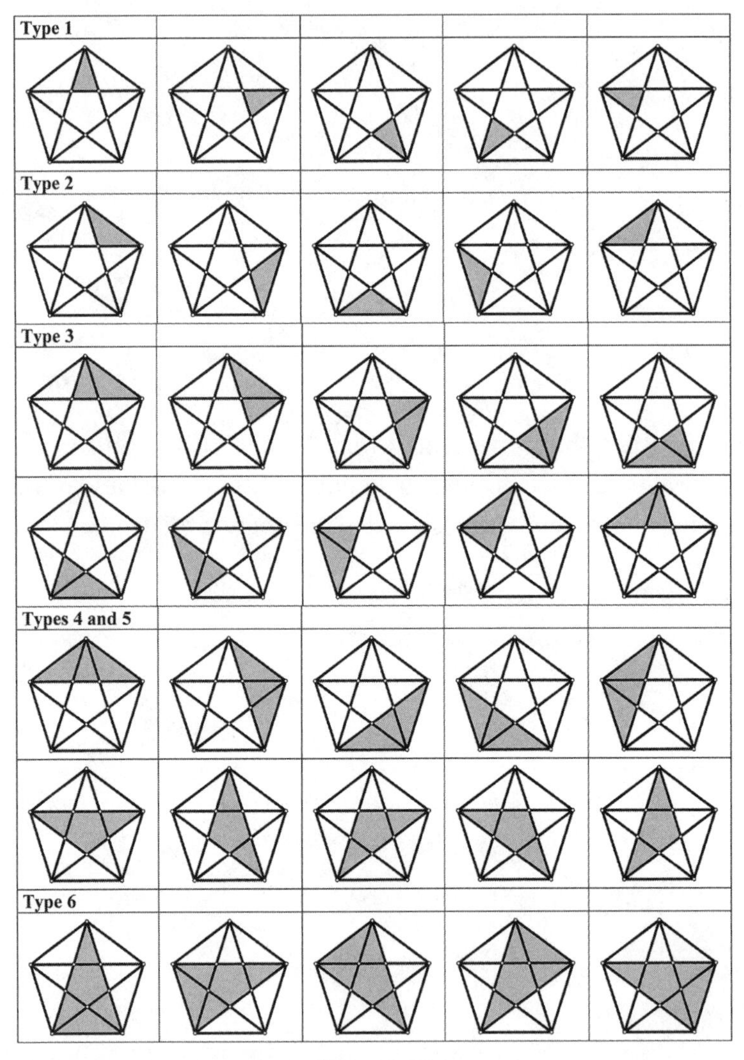

Figure 4.32.

How many different (noncongruent) triangles can be found in the regular pentagon? Since types 4 and 5 are congruent, there are only five different classes of noncongruent triangles in the regular pentagon.

HOW CAN A RIGHT ANGLE EQUAL AN OBTUSE ANGLE?

This geometric mistake points out a few properties that must hold and cannot be ignored. Furthermore, it shines a spotlight on a rarely recognized concept: the reflex angle. Follow along as we proceed to "prove" that a right angle can be equal to an obtuse angle (an angle that is greater than 90°).

We begin with a rectangle $ABCD$, where $FA = BA$, and where R is the midpoint of BC and N is the midpoint of CF (figure 4.33). We will now "prove that right angle CDA is equal to obtuse angle FAD.

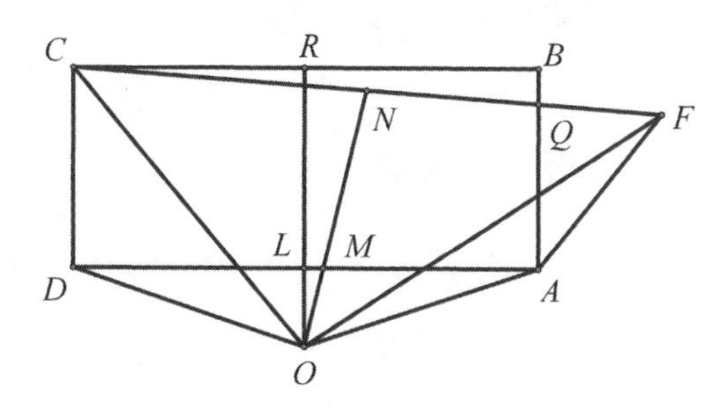

Figure 4.33.

To set up the proof, we first draw RL perpendicular to CB and draw MN perpendicular to CF. Then the rays RL and MN intersect at point O. If they did not intersect, then RL and MN would be parallel, and this would mean that CB would be parallel to or coincide with CF, which is impossible. To complete the diagram for our "proof," we draw the line segments DO, CO, FO, and AO.

We are now ready to embark on the "proof." Since RO is the perpendicular bisector of CB and AD, we know that $DO = AO$. Similarly, since NO is the perpendicular bisector of CF, we get $CO = FO$. Furthermore, since $FA = BA$, and $BA = CD$, we can conclude that $FA = CD$. This enables us to establish $\Delta CDO \cong \Delta FAO$ (SSS), so that $\angle ODC = \angle OAF$. We continue with $OD = OA$, which makes triangle AOD isosceles, and then the base angles ODA and OAD are equal. Now, $\angle ODC - \angle ODA = \angle OAF - \angle OAD$ or $\angle CDA = \angle FAD$. This says that a right angle is equal to an obtuse angle. There must be some mistake!

Clearly, there is nothing wrong with this "proof," however, if you use a ruler and compasses to reconstruct the diagram, it will look like figure 4.34.

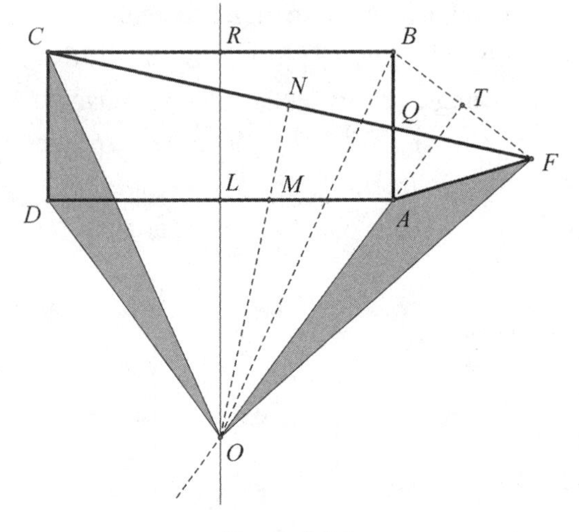

Figure 4.34.

As you will see, the mistake here rests with a reflex angle—one that is often not considered. For rectangle $ABCD$, the perpendicular bisector of AD will also be the perpendicular bisector of BC. Therefore, $OC = OB$, $OC = OF$, and then $OB = OF$. Since both points A and O are equidistant from the endpoints of BF, the line AO must be the perpendicular bisector of BF. This is where the fault lies; we must consider the reflex angle of angle BAO. Although the triangles are congruent, our ability to subtract the specific angles no longer exists. Thus, the difficulty with this "proof" lies in its dependence upon an incorrectly drawn diagram.

A MISTAKEN "PROOF" THAT EVERY ANGLE IS A RIGHT ANGLE

We begin this demonstration with quadrilateral $ABCD$, where $AB = CD$ and right angle $\angle BAD = \delta$ (see figure 4.35). We will keep $\angle ADC = \delta'$ to be of random measure, but we will show that it is actually a right angle. By showing this, we will have proved that any random angle is a right angle.

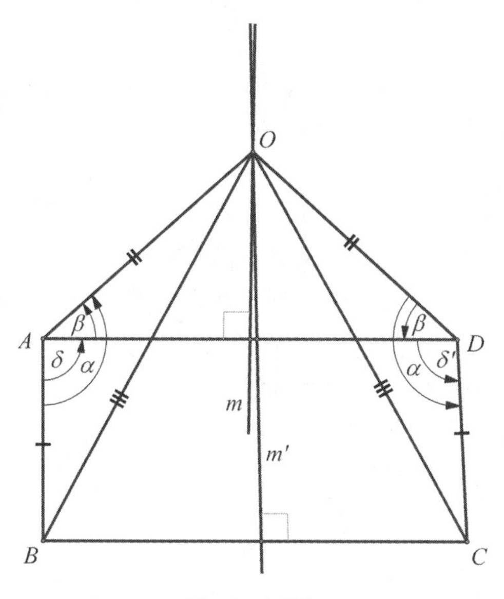

Figure 4.35.

We then construct m, the perpendicular bisector of AD, and m', the perpendicular bisector of BC. These perpendicular bisectors intersect at that point O. The point O is then equidistant from points A and D, as well as from points B and C. Therefore, $OA = OD$ and $OB = OC$. We can then conclude that $\Delta OAB \cong \Delta ODC$, and it follows that $\angle BAO = \angle ODC = \alpha$.

Since triangle OAD is isosceles, it follows that $\angle DAO = \angle ODA = \beta$.

Therefore, $\delta = \angle BAD = \angle BAO - \angle DAO = \alpha - \beta$, and $\delta' = \angle ADC = \angle ODC - \angle ODA = \alpha - \beta$.

And then follows that $\delta = \delta'$.

However, this result is silly. There must be a mistake somewhere. Let's revisit the original diagram.

In fact, the diagram presented in figure 4.35 tricked us and was intentionally false. The key error is the point where the two perpendicular bisectors meet, which must be further beyond the quadrilateral than what was indicated. The correct diagram would look like that shown in figure 4.36. We then have $\delta = \alpha - \beta$, however, $\delta' = 360° - \alpha - \beta$. This then destroys the mistaken "proof."

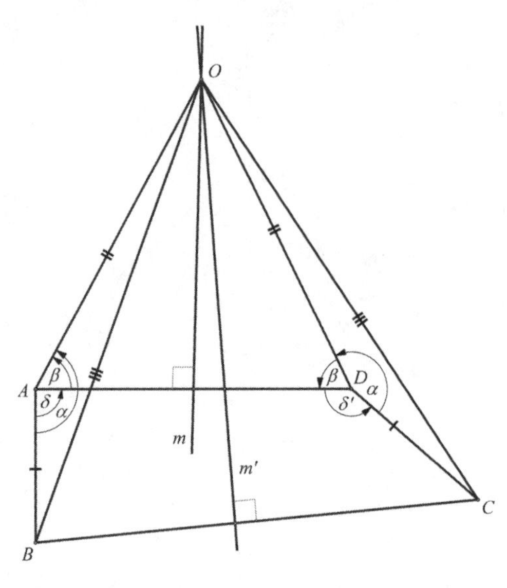

Figure 4.36.

HOW CAN 64 = 65?

We now have a mathematics mistake that was popularized by Charles Lutwidge Dodgson (1832–1898), who, under the pen name of Lewis Carroll, wrote *The Adventures of Alice in Wonderland.* In figure 4.37, we notice that the square on the left side has an area of $8 \cdot 8 = 64$ and is partitioned into two congruent trapezoids and two congruent right triangles. Yet when these four parts are placed into a different configuration (as shown on the right side of figure 4.37), we get a rectangle whose area is $5 \cdot 13 = 65$. How can $64 = 65$? There must be a mistake somewhere.

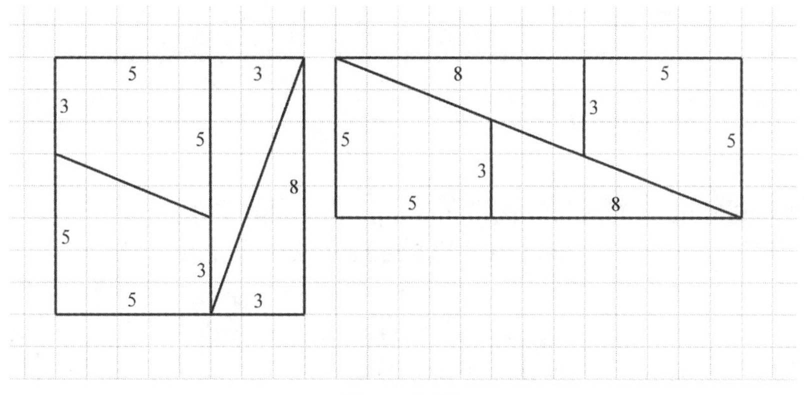

Figure 4.37.

When we correctly construct the rectangle formed by the four parts of the square, we find that there is an extra parallelogram in the figure—as shown, exaggerated in size, in figure 4.38.

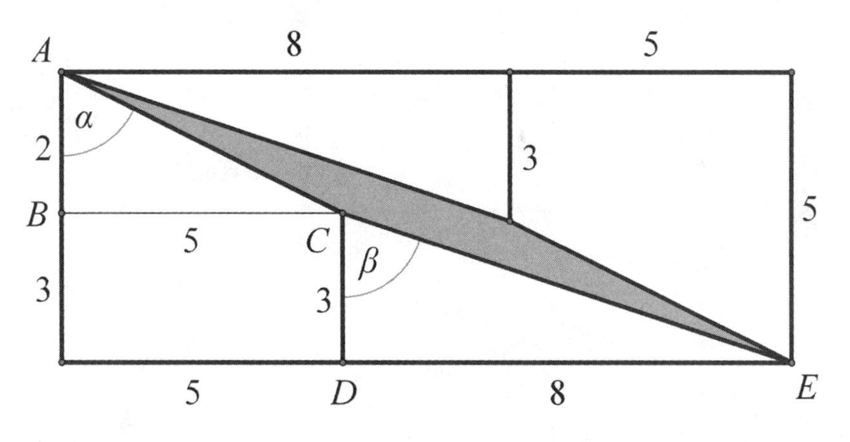

Figure 4.38.

This parallelogram (shaded) results from the fact that the angles marked α and β are not equal. Yet this is not easily noticeable at a glance in the original diagram!

Perhaps the easiest way to show this is to refer to the familiar tangent function. In triangle ABC, the $\tan \alpha = \frac{5}{2} = 2.5$, while the $\tan \beta = \frac{8}{3} \approx 2.667$. In order for the line segment ACE to be a straight line—preventing a parallelogram from being formed—the angles α and β would have to be equal. With different tangent values, this is not the case! Thus the mistake—one easily overlooked—has been exposed.[3]

ANOTHER MISTAKEN "PROOF" THAT SHOWS THAT TWO RANDOMLY DRAWN LINES IN A PLANE ARE ALWAYS PARALLEL

We begin this demonstration with the two randomly drawn lines, l_1 and l_2. We then construct two parallel lines *AD* and *BC* that intersect our two given lines l_1 and l_2. We complete our required diagram by drawing $EF \parallel AD$. The line *EF* intersects *BD* and *AC* in points *G* and *H*, respectively (see figure 4.39).

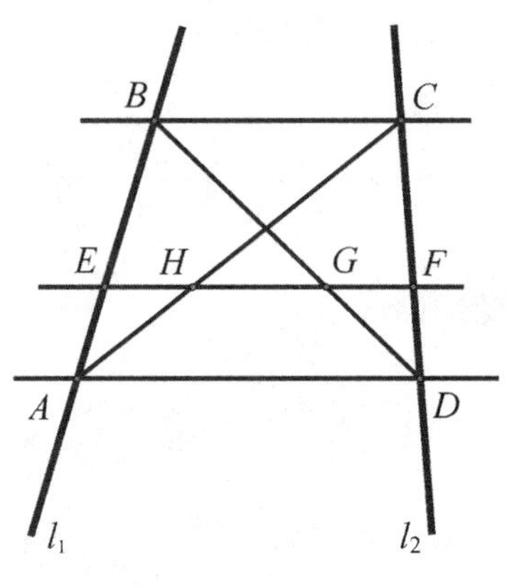

Figure 4.39.

The triangles *AEH* and *ABC* are similar, as are the triangles *HCF* and *ACD*. We, therefore, can establish the following proportions: $\frac{EH}{BC} = \frac{AH}{AC}$, and $\frac{HF}{AD} = \frac{HC}{AC}$.

When we add the two proportions, we get the following:

$$\frac{EH}{BC} + \frac{HF}{AD} = \frac{AH}{AC} + \frac{HC}{AC} = \frac{AH + HC}{AC} = \frac{AC}{AC} = 1, \text{ which is to say that } \frac{EH}{BC} + \frac{HF}{AD} = 1.$$

Analogously we can establish the similarity between the triangles *BGE* and *BDA* as well as triangles *BDC* and *GDF* and then get the following result:

$$\frac{EG}{AD} + \frac{GF}{BC} = 1.$$

Since the last two equations are equal to 1, we get:

$$\frac{EH}{BC} + \frac{HF}{AD} = \frac{EG}{AD} + \frac{GF}{BC}, \text{ or } \frac{HF}{AD} - \frac{HF}{AD} = \frac{GF}{BC} - \frac{EH}{BC}.$$

Therefore, $\frac{HF - EG}{AD} = \frac{GF - EH}{BC}$.

From the diagram, we find that $HF - EG = (EF - EH) - (EF - GF) = GF - EH$. This tells us that the numerators of the two equal fractions are equal. Consequently, the denominators must also be equal. Therefore, $AD = BC$. Since we began with $AD \parallel BC$, the quadrilateral $ABCD$ must be a parallelogram, and therefore, $AB \parallel CD$, or $l_1 \parallel l_2$.

Thus, we seem to have proved that two randomly drawn lines in the same plane are actually parallel. Clearly this is absurd, and so there must have been a mistake made in this demonstration.

Let's take another look at what we have just done. From figure 4.39 you can clearly see that $HF - EG = (HG + GF) - (EH + HG) = GF - EH$.

From the parallel lines in the diagram the following proportions follow immediately: $\frac{EH}{BC} = \frac{AE}{AB} = \frac{AH}{AC} = \frac{DF}{DC} = \frac{GF}{BC}$.

Since $BC \neq 0$, we then have $EH = GF$. Therefore, $GF - EH = 0$, and $HF - EG$ must also equal 0.

From the earlier equation, $\frac{HF - EG}{AD} = \frac{GF - EH}{BC}$. By substitution, we have the following: $\frac{0}{AD} = \frac{0}{BC}$.

This essentially tells us that we had no reason to state that $AD = BC$, since AD and BC can essentially take on any values to make this equation true. This explains where the mistake was made.

"PROVING" THAT A SCALENE TRIANGLE IS ISOSCELES, OR "PROVING" THAT ALL TRIANGLES ARE ISOSCELES: A MISTAKE?

Mistakes in geometry—also sometimes called fallacies—tend to come from faulty diagrams that result from a lack of definition. Yet, as we know, in ancient times some geometers discussed their geometric findings or relationships in the absence of a diagram. For example, in Euclid's work, the concept of "betweenness" was not considered. When not considering this concept, we can prove that any triangle is isosceles—that is, that a triangle that has three sides of different lengths actually has two sides that are equal. This sounds a bit strange, but we can demonstrate this "proof" and have the reader attempt to discover where the mistake lies—before we expose it.

We shall begin by drawing any scalene triangle (i.e., a triangle with no two sides of equal length) and then "prove" it is isosceles (i.e., a triangle with two sides of equal length). Consider the scalene triangle ABC, where we then draw the bisector of angle C and the perpendicular bisector of AB. From their point of intersection, G, draw perpendiculars to AC and CB, meeting them at points D and F, respectively.

We could now have four possibilities meeting the above description for various scalene triangles:

In figure 4.40, where CG and GE meet inside the triangle at point G.

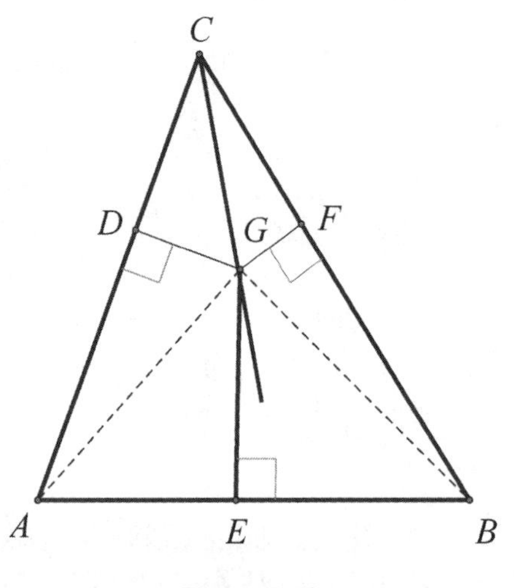

Figure 4.40.

In figure 4.41, where CG and GE meet on side AB. (Points E and G *coincide*.)

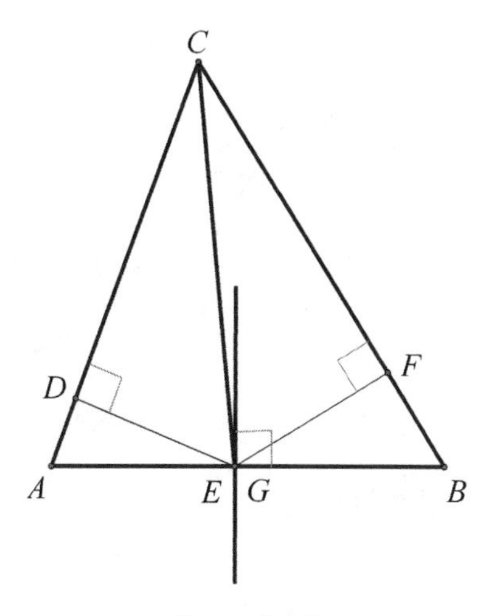

Figure 4.41.

In figure 4.42, where *CG* and *GE* meet outside the triangle (in *G*), but the perpendiculars *GD* and *GF* intersect the segments *AC* and *CB* (at points *D* and *F*, respectively).

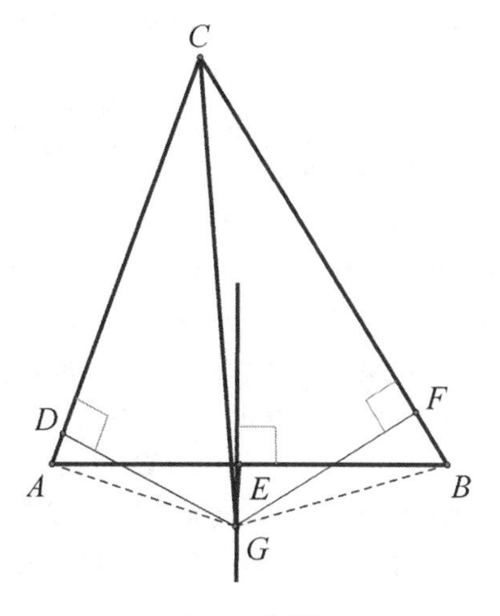

Figure 4.42.

In figure 4.43, where *CG* and *GE* meet outside the triangle, but the perpendiculars *GD* and *GF* intersect the extensions of the sides *AC* and *CB* outside the triangle (in points *D* and *F*, respectively).

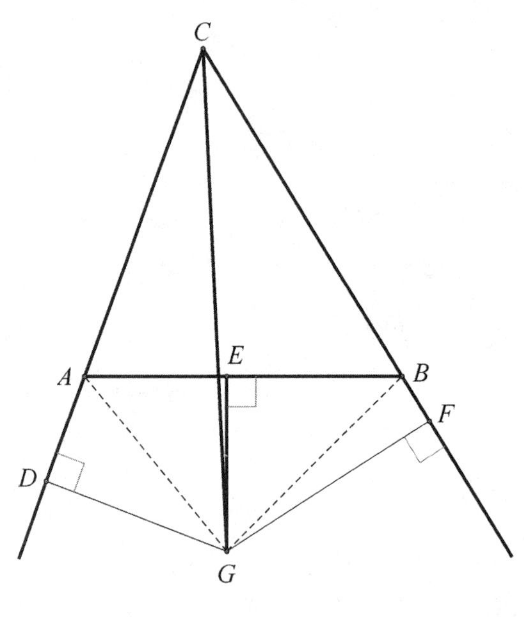

Figure 4.43.

The "proof" of the mistake or fallacy can be done with any of the above figures. Follow along and see if the mistake shows itself without reading further. We begin with a scalene triangle, ABC. We will now "prove" that $AC = BC$ (or that triangle ABC is isosceles).

As we have an angle bisector, we have $\angle ACG \cong \angle BCG$. We also have two right angles, such that $\angle CDG \cong \angle CFG$. This enables us to conclude that $\Delta CDG \cong \Delta CFG$ (SAA). Therefore, $DG = FG$ and $CD = CF$. Since a point on the perpendicular bisector (EG) of a line segment is equidistant from the endpoints of the line segment, $AG = BG$. Also, $\angle ADG$ and $\angle BFG$ are right angles. We then have $\Delta DAG \cong \Delta FBG$ (since they have respective hypotenuse and leg congruent). Therefore, $DA = FB$. It then follows that $AC = BC$ (by addition in figures 4.40, 4.41, and 4.42; and by subtraction in figure 4.43).

At this point you may feel quite disturbed. You may wonder where the error was committed, which permitted this mistake to occur. You might challenge the correctness of the figures. Well, by rigorous construction, you will find a subtle error in the figures. We will now divulge the mistake and see how it leads us to a better and more precise way of referring to geometric concepts.

First we can show that the point G *must* be outside the triangle. Then, when perpendiculars meet the sides of the triangle, one will meet a side *between* the vertices, while the other will not.

We can "blame" this mistake on Euclid's neglect of the concept of betweenness. However, the beauty of this particular mistake lies in the proof of this betweenness issue, which establishes the mistake.

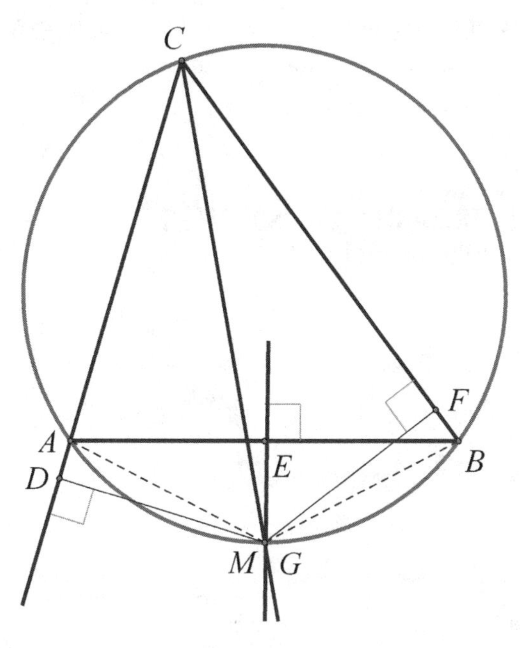

Figure 4.44.

Begin by considering the circumcircle of triangle ABC (figure 4.44). The bisector of angle ACB must contain the midpoint, M, of arc AB (since angles ACM and BCM are congruent inscribed angles). The perpendicular bisector of AB must bisect arc AB, and therefore pass through M. Thus, the bisector of angle ACB and the perpendicular bisector of AB intersect on the circumscribed circle, which is *outside* the triangle at M (or G). This eliminates the possibilities we used in figures 4.40 and 4.41.

Now consider the inscribed quadrilateral *ACBG*. Since the opposite angles of an inscribed (or cyclic) quadrilateral are supplementary, $\angle CAG + \angle CBG = 180°$. If angles *CAG* and *CBG* were right angles, then *CG* would be a diameter and triangle *ABC* would be isosceles. Therefore, since triangle *ABC* is scalene, angles *CAG* and *CBG* are not right angles. In this case, one must be acute and the other obtuse. Suppose angle *CBG* is acute and angle *CAG* is obtuse. Then in triangle *CBG*, the altitude on *CB* must be *inside* the triangle; while in obtuse triangle *CAG*, the altitude on *AC* must be *outside* the triangle. The fact that one and *only one* of the perpendiculars intersects a side of the triangle *between* the vertices destroys the fallacious "proof." This demonstration hinges on the definition of *betweenness*, a concept not available to Euclid.

"PROVING" ALL TRIANGLES ARE ISOSCELES: ANOTHER MISTAKEN PROOF!

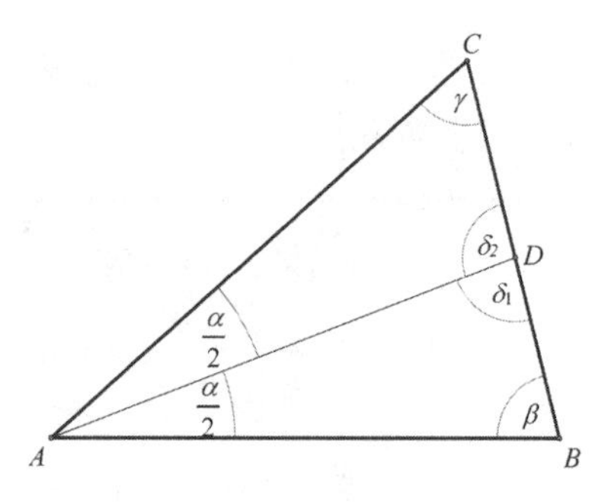

Figure 4.45.

We will give another "proof" that *every* triangle is isosceles. Again we must search for the mistake.

Here we begin with triangle *ABC*, where $AB \neq AC$, and we will "prove" that $AB = AC$ (see figure 4.45). We will require an auxiliary line, specifically the angle bisector *AD*, which will give us the relationship[4]: $\frac{CD}{AC} = \frac{BD}{AB}$.

For the exterior angle of triangle *ACD*, we have $\delta_1 = \angle ADB = \angle ACD + \angle CAD = \gamma + \frac{\alpha}{2}$. We can apply the law of sines to triangle *ABD* to get:

$$\frac{BD}{AB} = \frac{\sin\angle BAD}{\sin\angle ADB} = \frac{\sin\frac{1}{2}\angle BAC}{\sin\left(\angle ACD + \frac{1}{2}\angle BAC\right)} = \frac{\sin\frac{\alpha}{2}}{\sin\left(\gamma + \frac{\beta}{2}\right)}.$$

In a similar fashion, we apply the exterior angle property to triangle ABD to get $\delta_2 = \angle ADC = \angle ABD + \angle BAD = \beta + \frac{\alpha}{2}$. And again applying the law of sines, we get:

$$\frac{CD}{AC} = \frac{\sin\angle DAC}{\sin\angle ADC} = \frac{\sin\frac{1}{2}\angle BAC}{\sin\left(\angle ABC + \frac{1}{2}\angle BAC\right)} = \frac{\sin\frac{\alpha}{2}}{\sin\left(\beta + \frac{\alpha}{2}\right)}.$$

From our first relationship formed by the angle bisector, we can now equate these as:

$$\frac{\sin\frac{\alpha}{2}}{\sin\left(\gamma + \frac{\beta}{2}\right)} = \frac{\sin\frac{\alpha}{2}}{\sin\left(\beta + \frac{\alpha}{2}\right)}.$$

We know that $\sin\frac{\alpha}{2} \neq 0$. Therefore, $\sin\left(\gamma + \frac{\alpha}{2}\right) = \sin\left(\beta + \frac{\alpha}{2}\right)$, so that $\gamma + \frac{\alpha}{2} = \beta + \frac{\alpha}{2}$. Thus, we have $\gamma = \beta$ or $\angle ACD = \angle ABC$, and the triangle ABC is isosceles.

The error is hidden behind the sine function, which, at first glance, may not be too clear.

From $\sin\left(\gamma + \frac{\alpha}{2}\right) = \sin\left(\beta + \frac{\alpha}{2}\right)$, it follows that

$$\gamma + \frac{\alpha}{2} = (-1)^k\left(\beta + \frac{\alpha}{2}\right) + k\pi, \text{ or } \gamma = (-1)^k\beta + \frac{\alpha \cdot ((-1)^k - 1)}{2} + k\pi,$$

for all integer values of k.

A MISTAKEN PROOF THAT A TRIANGLE CAN HAVE TWO RIGHT ANGLES

The next geometric mistake is one that can truly upset an unsuspecting person. With two intersecting circles of different or the same size, we will draw the diameters from one of their points of intersection and then connect the other ends of the diameters, as shown in figure 4.46.

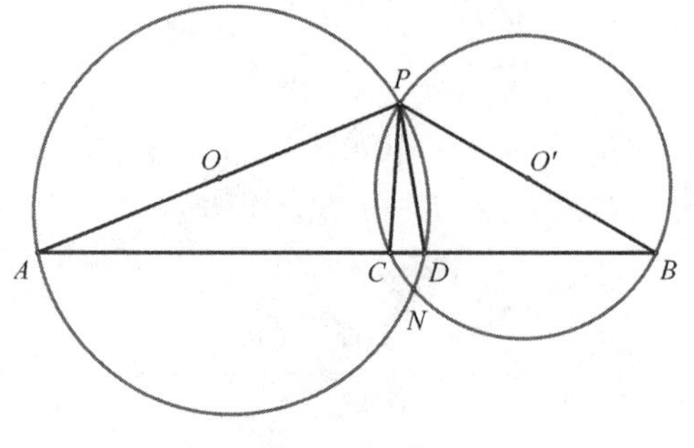

Figure 4.46.

In figure 4.46, where the endpoints of diameters AP and BP are connected by line AB, which intersects circle O at point D and circle O' at point C, we find that $\angle ADP$ is inscribed in semicircle PNA, and $\angle BCP$ is inscribed in semicircle PNB, thus making them both right angles. We then have a dilemma: triangle CPD has two right angles! This is impossible. Therefore, there must be a mistake somewhere in our work.

The concern with the omission of the concept of betweenness in Euclid's work could lead us to this dilemma. When we draw this figure correctly, we find that the angle CPD must equal 0, since a triangle cannot have more than 180°. That would make the triangle CPD nonexistent. Figure 4.47 shows the correct drawing of this situation.

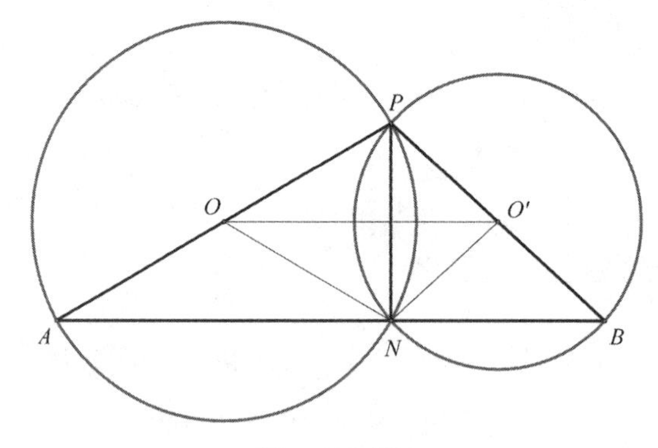

Figure 4.47.

In figure 4.47, we can easily show that $\Delta POO' \cong \Delta NOO'$, and then $\angle POO' = \angle NOO'$. Because $\angle PON = \angle A + \angle ANO$, and $\angle ANO = \angle NOO'$

(alternate-interior angles) we have $\angle POO' = \angle A$, and then $AN \parallel OO'$. The same argument can be made for circle O' to get $BN \parallel OO'$. Since each of the two line segments AN and BN are parallel to OO', they must in fact be one line ANB. This proves that the diagram in figure 4.47 is correct and the diagram in figure 4.46 is not.

THE OVERSIGHT OF A COMMON MISTAKE

There are times when our reasoning is faulty, resulting from an oversight of an unjustified assumption. Take, for example, the proof that the sum of the angles of a triangle is 180°.

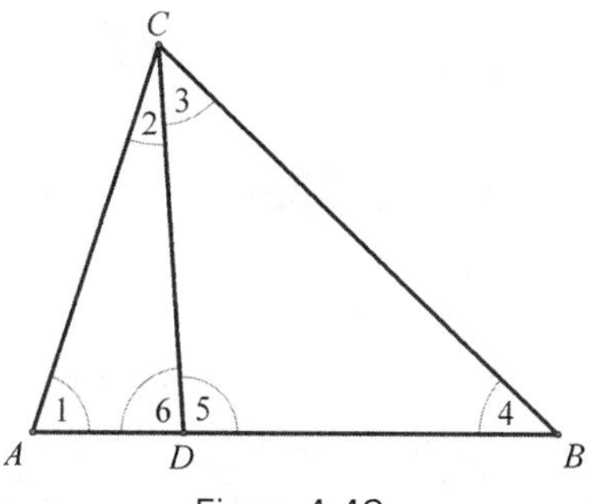

Figure 4.48.

In figure 4.48, we have triangle ABC with point D on AB. If we let x be the sum of the angles of triangles ACD and DCB, then $\angle 1 + \angle 2 + \angle 6 = x$, and $\angle 3 + \angle 4 + \angle 5 = x$.

By addition, we get $\angle 1 + \angle 2 + \angle 6 + \angle 3 + \angle 4 + \angle 5 = 2x$. Yet the sum $\angle 1 + \angle 2 + \angle 3 + \angle 4 = x$, since that is the sum of the angles of triangle ABC. However, since the angles 5 and 6 are supplementary, their sum is 180°. Therefore, we have $x + 180° = 2x$, and it follows that $x = 180°$. Hence, we have proved that the sum of the angles of a triangle is 180°. Wrong! There is a mistake in this "proof." We do not have the right to assume that the sum of the angles of a triangle is the same for all triangles, as we did at the outset of this "proof," when we assumed that the two triangles ACD and DCB each had and angle sum of x. The result is correct, but the proof was not complete and therefore was wrong!

TWO UNEQUAL LINES ARE ACTUALLY EQUAL

As we go through this "proof," see if you can spot the mistake. We will provide a clue: It has nothing to do with the diagram as was the case in previous examples of geometrical mistakes. We shall begin with triangle ABC and a line segment DE parallel to AB with endpoints on the other two sides as shown in figure 4.49.

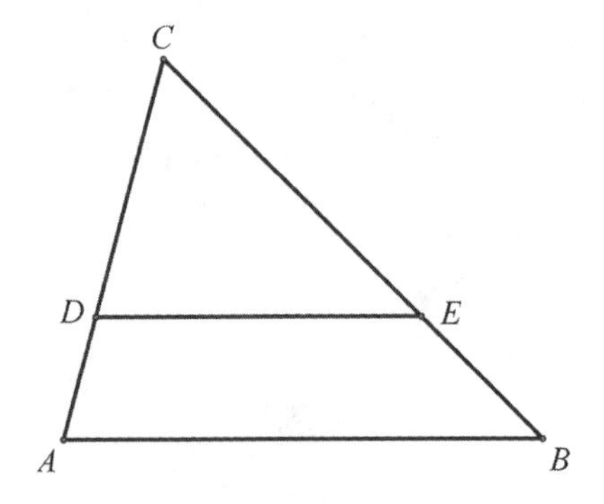

Figure 4.49.

Then we have $\Delta ABC \sim \Delta DEC$. Therefore, $\frac{AB}{DE} = \frac{AC}{DC}$, or $AB \cdot DC = DE \cdot AC$.

Now we will multiply both sides of this equation by $AB - DE$ to get $AB^2 \cdot DC - AB \cdot DC \cdot DE = AB \cdot DE \cdot AC - DE^2 \cdot AC$.

Our next step will be to add $AB \cdot DC \cdot DE$ and subtract $AB \cdot DE \cdot AC$ from both sides of the above equation to get $AB^2 \cdot DC - AB \cdot DE \cdot AC = AB \cdot DC \cdot DE - DE^2 \cdot AC$.

Factoring the common term on each side of the equation, $AB\,(AB \cdot DC - DE \cdot AC) = DE\,(AB \cdot DC - DE \cdot AC)$.

Now divide both sides by $AB \cdot DC - DE \cdot AC$, and we get $AB = DE$. This is absurd, since we can see that $AB > DE$. There was no error in the diagram, so where does the error lie? Yes, we divided by zero—recall the forbidden division! This occurred when we divided both sides of the equation above by $AB \cdot DC - DE \cdot AC$, which is equal to zero, since $AB \cdot DC = DE \cdot AC$. One must be aware that there are times—such as this—when an algebraic mistake creates a geometric absurdity.

EVERY EXTERIOR ANGLE OF THE TRIANGLE IS EQUAL TO ONE OF ITS REMOTE INTERIOR ANGLES

We begin with the triangle *ABC* shown in figure 4.50, and we would like to demonstrate that the angles δ and α are equal.

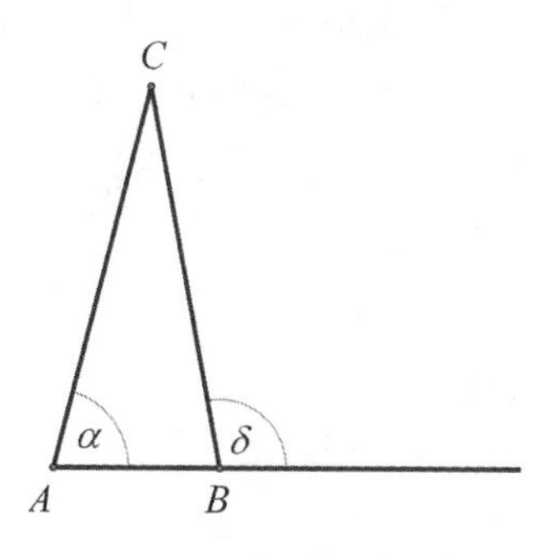

Figure 4.50.

We now refer to figure 4.51, where we have quadrilateral *APQC* so constructed that $\angle CAP + \angle CQP = \alpha + \varepsilon = 180°$. We then construct a circle through three points *C*, *P*, and *Q*. We will call the point where the line *AP* intersects the circle a second time point *B*. By drawing *BC* we have created a cyclic quadrilateral (i.e., one that can be inscribed in a circle) *BPQC*, where the following is true:

$$\angle CQP + \angle CBP = \varepsilon + \delta = \angle BCQ + \angle BPQ = 180°.$$

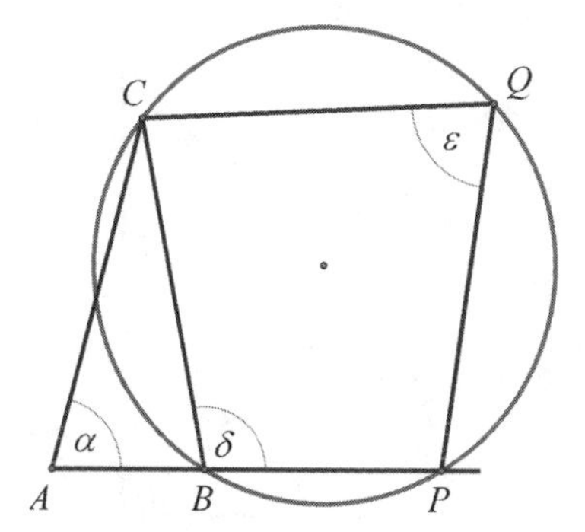

Figure 4.51.

However, at the outset we had drawn $\angle CAP + \angle CQP = \alpha + \varepsilon = 180°$, so that we can now conclude that $\angle CAP = \angle CBP$, which is to say that $\alpha = \delta$. Something must be wrong. Where does the mistake lie?

If quadrilateral $APQC$ has the property that $\angle CAP + \angle CQP = \alpha + \varepsilon = 180°$ and that the vertices C, P, and Q lie on the same circle, then the quadrilateral $APQC$ must also be cyclic, which implies that the point A must also lie on the circle. This implies that the two points A and B must be identical. In that case, the triangle ABC cannot exist. Thus, the mistake here has been revealed.

TWO NONPARALLEL LINES IN A PLANE THAT DO NOT INTERSECT: A PARADOX

We can also "prove" that if exactly one of two nonparallel lines is perpendicular to a third line, then the two nonparallel lines will not intersect. This paradox is attributed to Proclus Lycaeus (412–485).

There must be a mistake here since the only time the two lines will not intersect is if they are parallel, which is not the case here. So follow along and see if you can find the mistake. In figure 4.52, we have $PB \perp AB$, while QA is not perpendicular to AB. Now we will "show" that the nonparallel lines PB and QA cannot intersect.

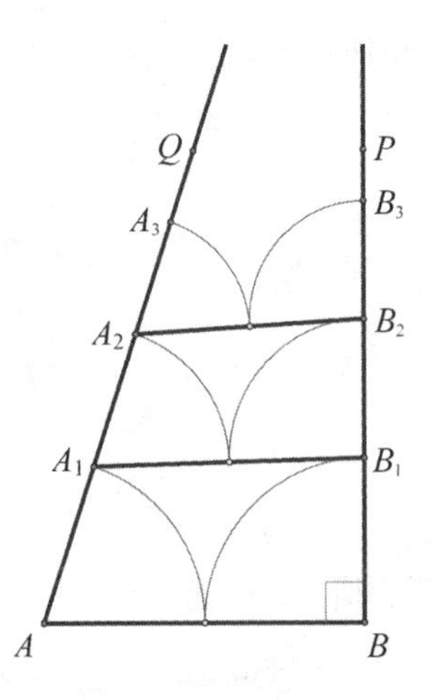

Figure 4.52.

We begin by finding the midpoint of AB. Then, mark off $AA_1 = \frac{1}{2}AB$, and $BB_1 = \frac{1}{2}AB$. The lines AQ and PB will not intersect anywhere along AA_1 or BB_1. If they did intersect there at, say, a point R, then there would be a triangle ARB, where the sum $AR + RB < AB$, which is impossible. We now consider the segment A_1B_1 and repeat the previous process so that we get $A_2A_3 = B_2B_3 = \frac{1}{2}A_2B_2$. As before the lines A_1A_2 and B_1B_2 cannot intersect, and A_2 cannot coincide with B_2. We continue this process, bisecting A_2B_2 and marking off $A_2A_3 = B_2B_3 = \frac{1}{2}A_2B_2$. This process continues indefinitely and we know that A_n will never coincide with B_n, since we would then have a right triangle where the hypotenuse AA_n would equal BB_n—clearly impossible! Therefore, at no step along this unending process will the oblique line intersect the perpendicular line. This is nonsensical! So where is the mistake?

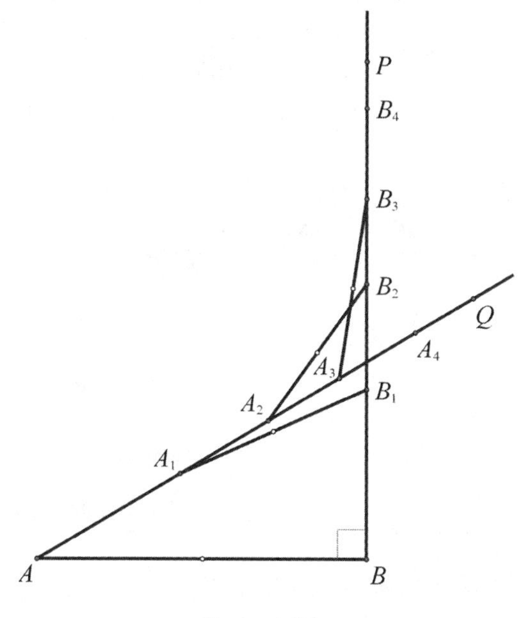

Figure 4.53.

Let's consider the two lines AQ and BP intersecting as shown in figure 4.53. Again we will denote the similarly constructed segments along AQ as we have before: $AB_1, A_1A_2, A_2A_3, \ldots$, and similarly the segments along BP as $BB_1, B_1B_2, B_2B_3, \ldots$. We know that the marking-off of these segments along the two lines can continue indefinitely. Furthermore, segments with the same indices will not intersect. That is, for example, A_1A_2 and B_1B_2 will not intersect. However, segments with different indices can intersect. For example, in figure 4.53, A_3A_4 intersects B_1B_2. The mistake with our "proof" was to rest our argument on the idea that only certain segments—those with same indices—will not intersect; but that doesn't mean that other segments cannot intersect. This mistake is based on a limited form of reasoning.

A MISTAKE RIGHT FROM THE START

Here we are faced with a simple problem from elementary geometry—one that could easily be found in a typical textbook. We are given a right triangle ABC (see figure 4.54) with hypotenuse of length $c = 4$, and one leg $b = \sqrt{12}$. Also $\angle BAC = \alpha = 40°$. We are being asked to find the length of side a.

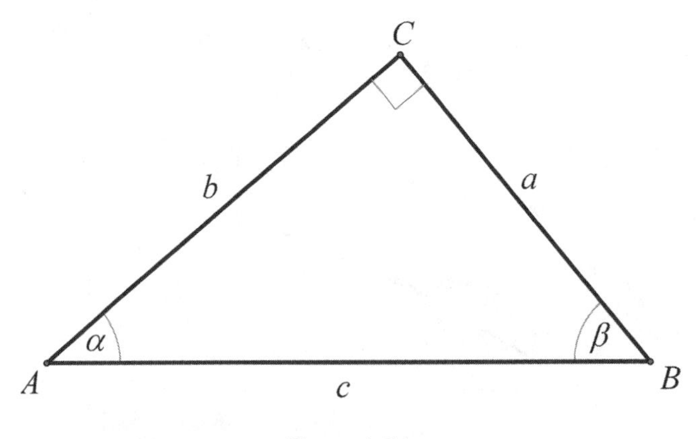

Figure 4.54.

We can easily find the measure of the third angle of the triangle as: $\beta = \angle ABC = 180° - \angle BAC - \angle BCA = 180° - \alpha - \gamma = 180° - 40° - 90° = 50°$.

Taking $\sin\alpha = \frac{BC}{AB} = \frac{a}{c}$, we get $a = c \cdot \sin\alpha = 4 \cdot \sin 40° \approx 4 \cdot 0.6428 = 2.5712$.

So far everything seems fine. Now comes the conundrum. We can also approach the problem in the following fashion:

Taking $\tan\alpha = \frac{BC}{AC} = \frac{a}{b}$, we get $a = b \cdot \tan\alpha = \sqrt{12} \cdot \tan 40° \approx 3.4641 \cdot 0.8391 = 2.9067$.

To find the third angle, we can take $\tan\beta = \frac{AC}{BC} = \frac{b}{a} \approx \frac{\sqrt{12}}{2.9067} \approx 1.1918$, whereby $\beta \approx 50.0001°$.

Now looking back at our two methods of solution, our measure of the angle β is essentially the same for both procedures. However, we arrived at two different values of a. Namely 2.5712 and 2.9067. How can this be? There must be a mistake somewhere. Yet both methods of solution were quite correct. Here the error lies in the fact that there was a mistake in the statement of the original problem.

If we try to construct this original triangle, given the following parts: $\alpha = \angle BAC = 40°$, $\gamma = \angle ACB = 90°$, $b = \sqrt{12}$ and $c = 4$, we would end up with a partial triangle as shown in figure 4.55.

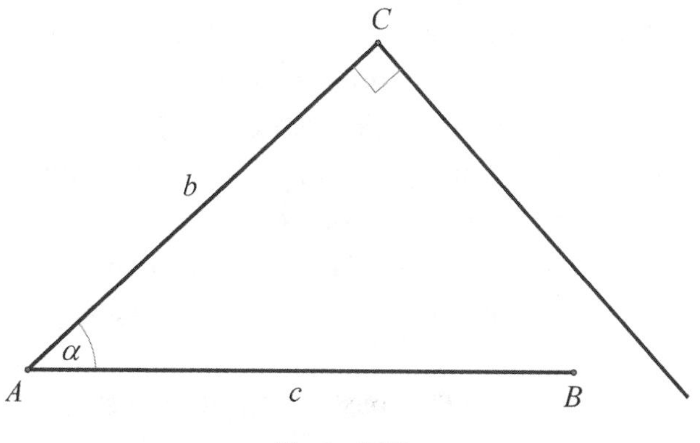

Figure 4.55.

In effect such a triangle cannot exist, thus the mistake here was in the original problem, that is, being given information about a triangle that cannot exist. Here we had an example of where the mistake was to accept given information that was faulty.

MISTAKES CAN ALSO BE MADE WITH DYNAMIC GEOMETRY PROGRAMS

With the proliferation of dynamic geometry drawing programs there is a tendency to ignore the time-tested—and correct—modes of construction. Take, for example, the construction of the inscribed circle of triangle ABC. Recall that the center of the inscribed circle is the point of intersection of the angle bisectors. The first step that one would take to begin such a construction would be to locate this center point.

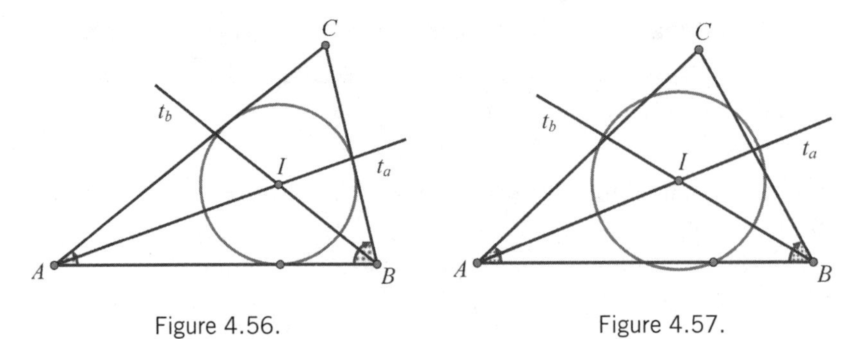

Figure 4.56.

Figure 4.57.

In figure 4.56, we see this done with the intersection of angle bisectors t_a and t_b, which locate the center I. Here is where a common mistake is made. The person doing the construction, in order to accomplish the task quickly, using (for example) Geometer's Sketchpad, will use the compass tool and draw a circle with the center at I and gradually pull out the circle until it touches (tangent) one of the sides, as shown in figure 4.56. However, this is clearly seen as a faulty construction when the original triangle is even just slightly distorted, as shown in figure 4.57, where the tangency is destroyed.

The correct way to do the construction of the inscribed circle is to do it as one would with straightedge and compass, by constructing a perpendicular line DI (= EI = FI) to one of the sides and then use that length as the radius to draw the circle. (See figure 4.58.) In other words, if one tries to create a shortcut with the dynamic geometry software, deviating from the traditional ruler-compass constructions, a mistake will generally be committed.

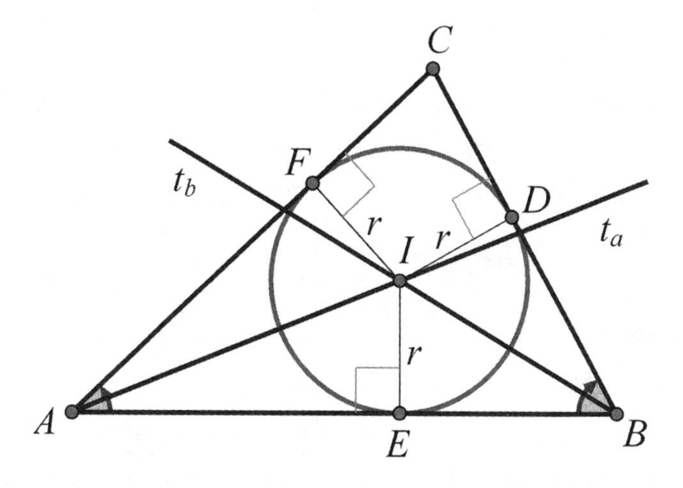

Figure 4.58.

THE MISTAKEN DIAGRAM LEADS TO A FAULTY CONCLUSION

There are times when we can draw a diagram of a situation that appears to be correct for some cases, but not for all. The proof of the general situation is then wracked with a mistake. We have such a case now. Is it true that a rectangle inscribed in a square is always a square? Please note that a rectangle is said to be inscribed in another rectangle (or square) when each vertex of the inner rectangle is on a unique side of the outer rectangle.

In figure 4.59, we have square *ABCD*, and rectangle *PRMN* inscribed in it. We will then draw perpendiculars *PQ* and *RS* to sides *BC* and *DC*, respectively, thereby forming two shaded triangles, ΔPQM and ΔRSN. Since the diagonals of a rectangle are equal, for rectangle *PRMN*, the diagonals *PM* and *RN* are equal. We also know that perpendicular segments *PQ* and *RS* are each equal to a side of the square, and so they, too, are equal to each other. Consequently, $\Delta PQM \cong \Delta RSN$, and then $\angle QMP = \angle SNR$. Since $\angle OMC + \angle QMP = 180°$, we have $\angle OMC + \angle SNR = 180°$. Therefore, in quadrilateral *NOMC*, we must have $\angle NOM + \angle NCM = 180°$. But $\angle NCM$ is a right angle; therefore, $\angle NOM$ must also be a right angle. Therefore, rectangle *PRMN* must be a square, since the diagonals are perpendicular.

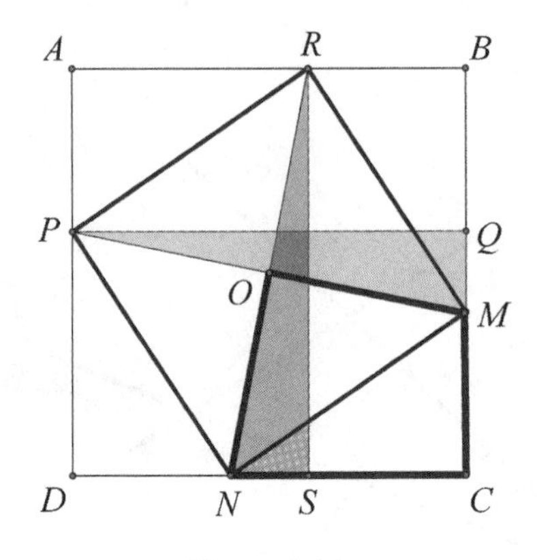

Figure 4.59.

So where is the mistake? If we look at figure 4.60, we can see a rectangle inscribed in a square that is clearly *not* a square. We placed it so that $AR = AP = CM = CN$, but not so that these equal segments would be equal to one-half the side of the original outer square. We can see that the orientation of the two congruent (shaded) triangles is different from that in figure 4.59. Whereas in the figure 4.59 the angles *OMC* and *ONC* were supplementary, they are now (figure 4.60) equal and *not* supplementary. Thus we cannot establish perpendicularity for the diagonals of the rectangle *PRMN* as we did before. For some rectangles inscribed in a square, we can prove the rectangle is a square, but not for all cases; therefore, there is a mistake in the original (general) proof.

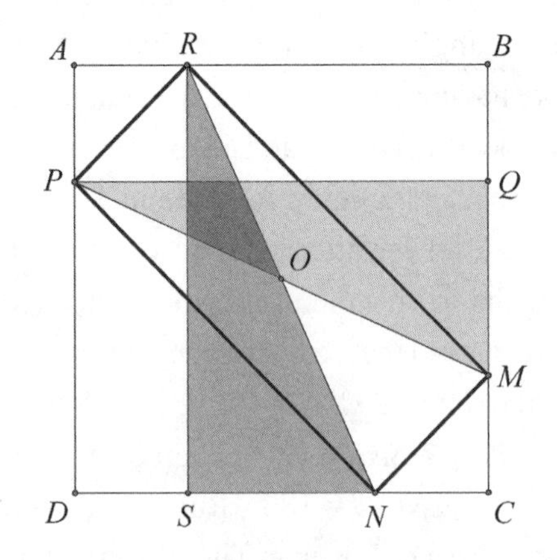

Figure 4.60.

From this mistaken proof of the erroneous statement, which we now cleared up, we can establish two true statements about this situation (figure 4.60):

1. If a rectangle is inscribed in a square so that one of the sides of the rectangle is *not* parallel to either of the diagonals of the square, then the rectangle is a square.
2. If a rectangle with unequal sides is inscribed in a square, then the sides of the rectangle must be parallel to the diagonals of the square.

A TRAPEZOID THE LENGTHS OF WHOSE BASES HAVE A SUM OF ZERO!

The mistake in this proof is very subtle and perhaps a bit difficult to find, but we should consider it, and then reveal the mistake. We begin with a trapezoid *ABCD*, and extend the bases (as shown in figure 4.61) to points *E* and *F*. The segment lengths are marked in the figure.

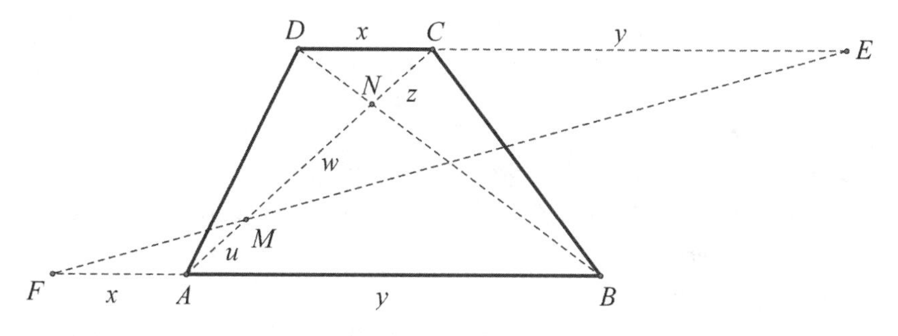

Figure 4.61.

From the parallel lines (the bases of the trapezoid) we get some similar triangles:

$\Delta CEM \sim \Delta AFM$, which gives us the proportion: $\frac{AF}{CE} = \frac{AM}{CM}$, or $\frac{x}{y} = \frac{u}{w+z}$, and

$\Delta ABN \sim \Delta CDN$, which gives us the proportion: $\frac{CD}{AB} = \frac{CN}{AN}$, or $\frac{x}{y} = \frac{z}{u+w}$.

Therefore, $\frac{u}{w+z} = \frac{z}{u+w}$. There is a convenient (legitimate) operation on proportions that allows subtraction across the numerators and denominators, such as if $\frac{a}{b} = \frac{c}{d}$, then $\frac{a}{b} = \frac{a-c}{b-d}$.

We can now apply that process to the above proportions, so that

$$\frac{x}{y} = \frac{u-z}{(w+z)-(u+w)} = \frac{u-z}{z-u} = \frac{-(z-u)}{z-u} = -1.$$

This would lead us to conclude that $x = -y$, or that $x + y = 0$. But how can the sum of the bases of the trapezoid be zero? Somewhere in this process there must have been a mistake. Let's look back over our development of this conclusion.

Suppose we solve the earlier two equations, $\frac{x}{y} = \frac{u}{w+z}$ and $\frac{x}{y} = \frac{z}{u+w}$, for u and z in terms of x, y, and w to get for the first equation:

$$yu = x(w + z)$$
$$yu = xw + xz$$
$$xz - yu = -xw.$$

For the second equation, we get:

$$yz = x(u + w)$$
$$yz = xu + xw$$
$$yz - xu = xw.$$

Now by adding these two newly obtained equations, we get: $(xz - yu) + (yz - xu) = (xz + yz) - (yu + xu) = 0$. Then, by rearranging terms, we can write this equation as $z(x + y) - u(x + y) = 0$. By factoring the $(x + y)$ term, we have: $(x + y)(z - u) = 0$.

As usual, if either factor is zero, then the equation will be satisfied. In our "proof" (above) we neglected the possibility that $(z - u)$ might be equal to zero, and just assumed that $(x + y)$ was equal to zero. However, logic tells us that $(x + y)$, which is the sum of the bases of the trapezoid, is clearly not equal to zero, therefore $(z - u)$ must be equal to zero. With that, the fraction above, $\frac{-(z-u)}{z-u}$, becomes $\frac{0}{0}$, which is meaningless!

CONSTRUCTING THE INSCRIBED CIRCLE OF A KITE: TRAPPED IN A MISTAKE[5]

A kite is a quadrilateral with two pairs of opposite adjacent sides of equal length; or, one might say that a kite is formed by two isosceles triangles sharing a common base and not overlapping. In figure 4.62, quadrilateral *ABCD* is a kite. To find the center *I* of the inscribed circle of this quadrilateral, we merely connect the midpoints of opposite sides. This point of intersection is the center of the inscribed circle. To construct the circle, we still need to find the length of the radius. This is found by taking the perpendicular length from the center to one of the sides. In this case, side *IP* is drawn perpendicular to *AB*.

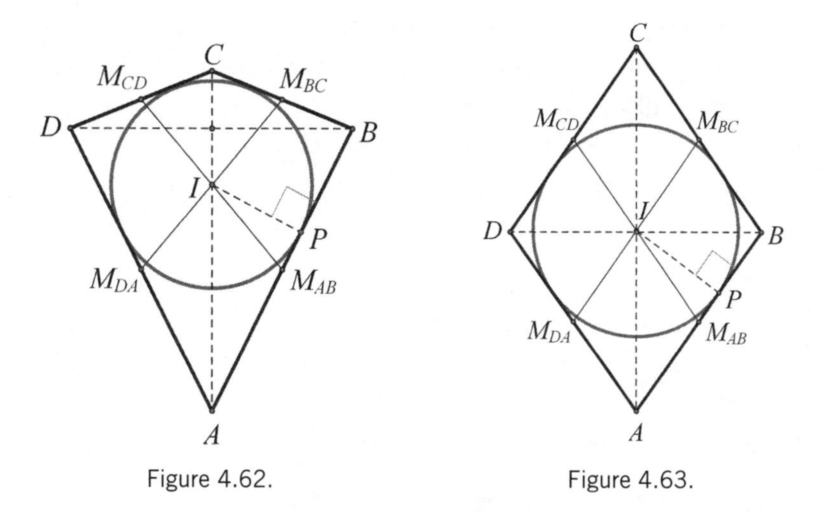

Figure 4.62.

Figure 4.63.

This rather elegant construction seemed to work for the kite shown in figure 4.62; yet it may not work for all kites. It will, however, work for kites that are also a rhombus (see figure 4.63). Of course, since a square is a special form of rhombus, it will also work for a square, where the midpoints of the sides are also the points of tangency of the inscribed circle.

Unfortunately, this elegant construction, which worked in figure 4.62, is sometimes thought to be applicable to all kites. This is not the case, as you can see in figures 4.64 and 4.65.

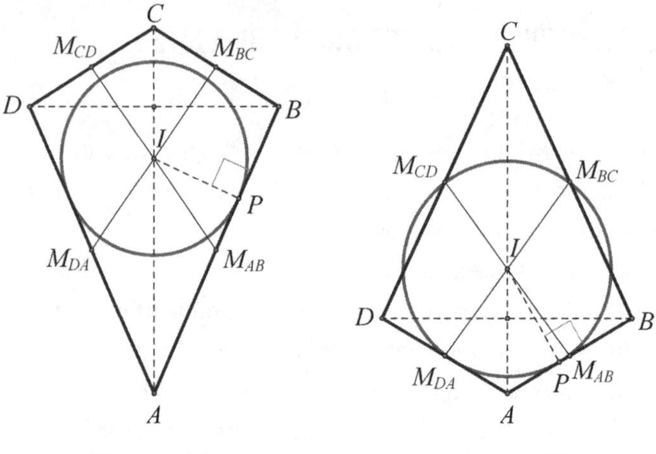

Figure 4.64.

Figure 4.65.

However, all (convex) kites do have an inscribed circle, that is, a circle that is tangent to all four sides. This true center of the inscribed circle is found by taking the angle bisectors of each of the angles of the kite. In figures 4.66 and 4.67 we shall consider these inscribed circles, whose center is I and where the intersection of the lines joining the midpoints of the sides is I'.

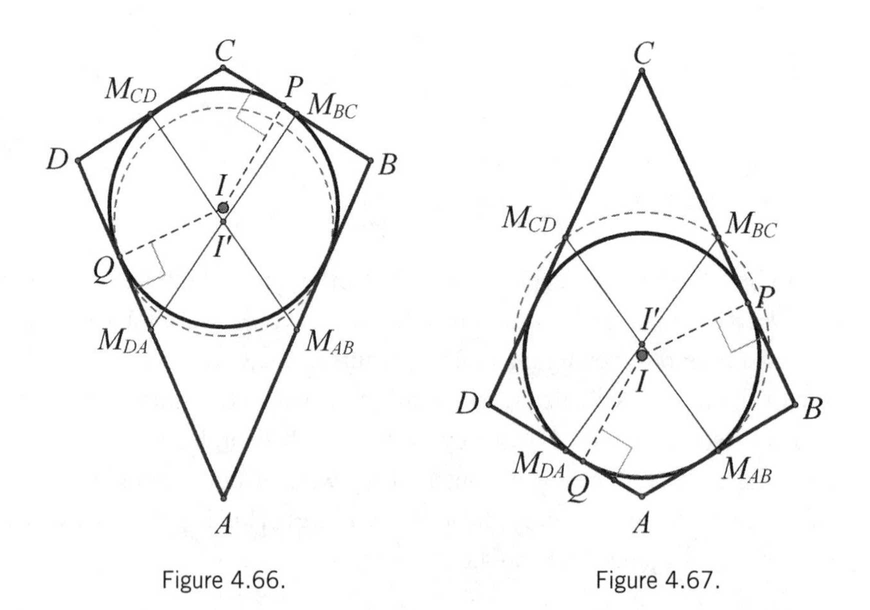

Figure 4.66.

Figure 4.67.

Now you may be wondering, are there any kites, which are not rhombuses, where the center of the inscribed circle can be found by taking the intersection of the lines joining the midpoint of the opposite sides of the kite? That is to say, how can we avoid the mistake of generalizing the initial construction to all forms of kites? It turns out that those kites that have the following characteristics will have the center of their inscribed circle at the point of intersection of the lines joining the midpoints of the opposite sides.

In figure 4.68, we have a kite with one pair of opposite vertices that lie on an ellipse, and the other pair of opposite vertices lie on the foci of this ellipse. The construction lines are shown in this figure.

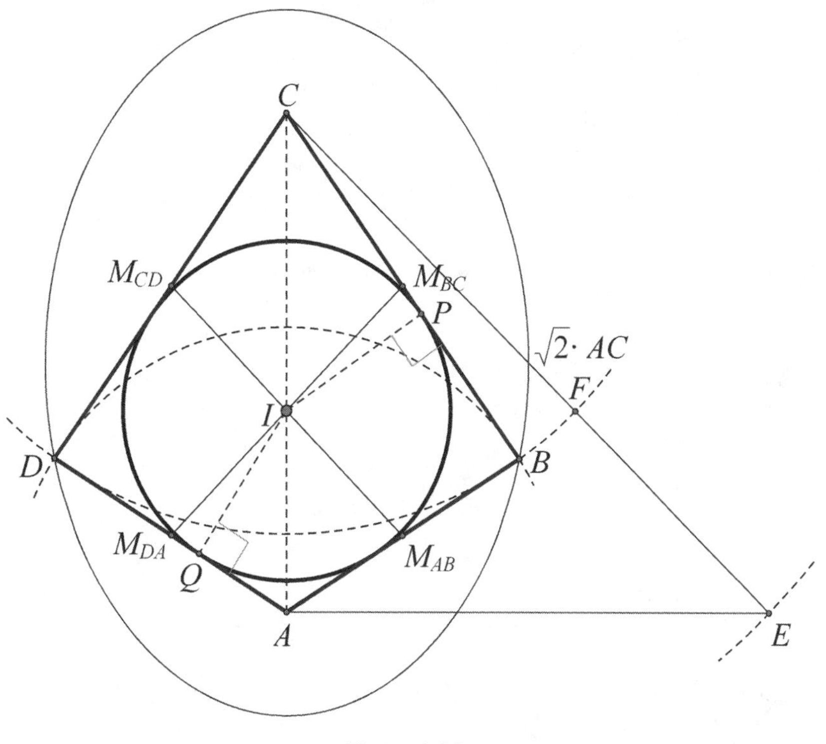

Figure 4.68.

The mistake of generalizing an elegant construction can be alluring, but it must be justifiable. Once we uncover its limitations, the mistaken generalization leads us to a full appreciation of the properties of a kite and shows its relationship to the ellipse.

ANY POINT IN THE INTERIOR OF A CIRCLE IS ALSO ON THE CIRCLE

Let's consider the conflicting statement that any point in the interior of a circle is also on the circle. It sounds ridiculous, but we can provide a "proof" of this statement. There must be a mistake, or else we are in a logical dilemma.

We shall begin our "proof" with a circle O, whose radius is r (see figure 4.69). We will then let A be any point in the *interior* of the circle distinct from O, and "prove" that the point A is actually *on* the circle.

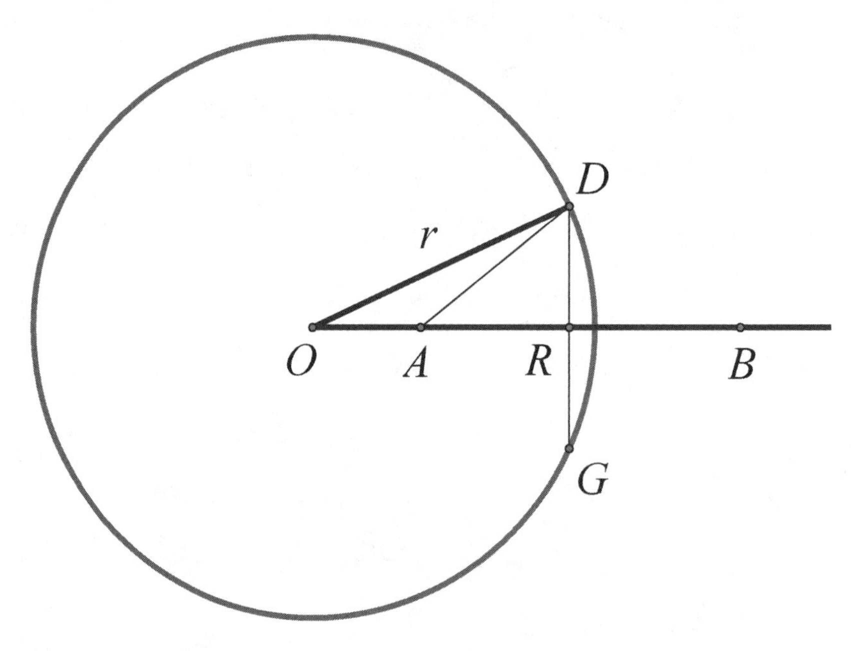

Figure 4.69.

We will set up our diagram as follows: let B be on the extension of OA through A such that $OA \cdot OB = OD^2 = r^2$. (Clearly OB is greater than r, since OA is less than r.) The perpendicular bisector of AB meets the circle in points D and G, where R is the midpoint of AB. We now have $OA = OR - RA$ and $OB = OR + RB = OR + RA$.

Therefore, $r^2 = OA \cdot OB = (OR - RA)(OR + RA)$, or $r^2 = OR^2 - RA^2$.

However, applying the Pythagorean theorem to triangle ORD, we get

$OR^2 = r^2 - DR^2$; and applying it once again to triangle ADR gives us $RA^2 = AD^2 - DR^2$. Therefore, since $r^2 = OR^2 - RA^2$, we get $r^2 = (r^2 - DR^2) - (AD^2 - DR^2)$, which reduces to $r^2 = r^2 - AD^2$. This would imply that $AD^2 = 0$; putting it another way, A coincides with D, and thus lies on the circle. That is, the point A inside the circle has been proved to be on the circle. There must be a mistake somewhere!

The fallacy in this proof lies in the fact that we drew an auxiliary line DRG with *two* conditions—that it is the perpendicular bisector of AB and that it intersects the circle. Actually, all points on the perpendicular bisector of AB lie in the exterior of the circle and therefore cannot intersect the circle.

Follow along with the algebraic process:

$$r^2 = OA \cdot OB$$
$$r^2 = OA(OA + AB)$$
$$r^2 = OA^2 + OA \cdot AB. \qquad (1)$$

The "proof" assumes that $OA + \frac{AB}{2} < r$. By multiplying both sides of the inequality by 2, we get: $2 \cdot OA + AB < 2r$.

By squaring both sides of the inequality, we have:

$$4 \cdot OA^2 + 4 \cdot OA \cdot AB + AB^2 < 4r^2. \qquad (2)$$

By substituting four times equation (1), which is $4r^2 = 4OA^2 + 4OA \cdot AB$, into equation (2), we get $4r^2 + AB^2 < 4r^2$, or $AB^2 < 0$, which is impossible.

The mistake here alerts us to the care that must be taken when allowing points to take on more properties than are possible. That is, when drawing auxiliary lines, we must take care that they use *one* condition only.

DO ALL CIRCLES HAVE THE SAME CIRCUMFERENCE?

Sometimes physical observations are very difficult to explain and can even be paradoxical. For example, we know that when a circle rolls on a line and makes one complete revolution, then it has traveled the distance equal to the length of its circumference. In figure 4.70, when the larger circle travels from point *A* to point *B*, it will have traveled the distance *AB*, which is equal to the circumference of the larger circle. When you consider the two concentric circles rolling, whose circumferences are not equal, we wonder how the smaller circle will have traveled one large-circle-circumference length at the same time as the larger circle traveled a longer distance. This may be seen in figure 4.70, where *AB* is equal to *CD*. In other words, the small circle and larger circle have the same circumference. This paradox dates back to Aristotle (384–322 BCE). How is this possible? Where is the mistake?

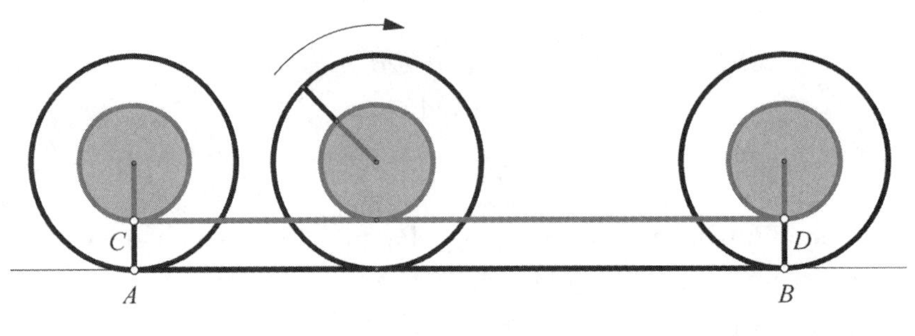

Figure 4.70.

If we observe a fixed point on each of the two circles during this rolling exercise—in this case, points *A* and *C*—we notice that the points travel on a cycloid path, as shown in figure 4.71.

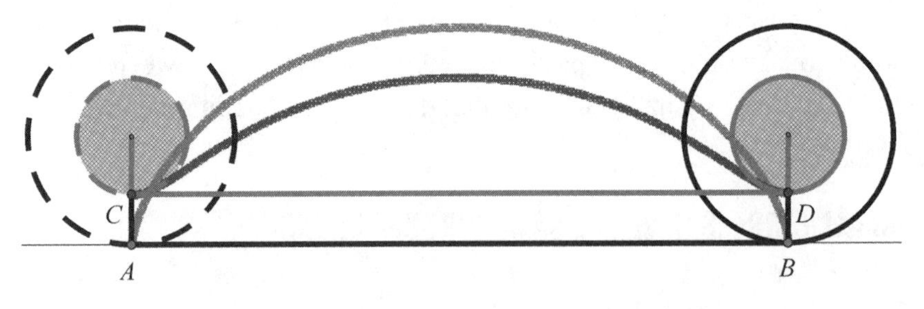

Figure 4.71.

With additional revolutions, this becomes even more clear, as you can see in figure 4.72.

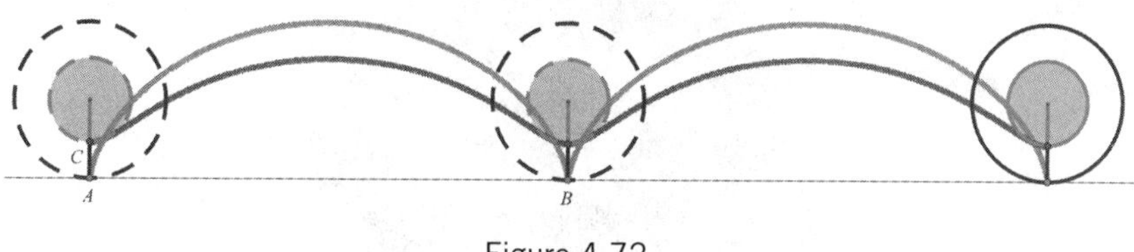

Figure 4.72.

The curves describe the paths of points A and C for two complete revolutions of the circles. Yet the paths along which they travel are not equal to the circumference of the circle on which they lie. The distance from A to B along the straight line is equal to $2\pi R$, where R is the length of the radius of the larger circle. We can clearly see that the cycloid curve between points A and B is longer than the circumference of the larger circle. The length of the cycloid of points on each of the two circles is dependent on the circle's circumference. Interestingly enough, the length of the cycloid can be an integer, if the radius is also an integer.[6]

The mistake that all circles have the same conference rests with the assumption that both circles simultaneously roll. The fallacy that we face here is actually not geometric; it is one of mechanics. Only one of the circles can roll at a time. If the larger circle rolls, then the smaller circle slides along. Were the smaller circle to roll, then the larger circle would slide somewhat backward. The mistake here is to not have recognized that the wheels are not rolling together. While one wheel rolls, the other wheel slides along. Thus, the error is a mechanical one.

FURTHER DECEPTION OF ROLLING CIRCLES

We begin by placing two touching coins on a table. One of the coins will remain stationary, while the other coin will rotate around the first coin without any slippage until it reaches the initial point—in other words, when the center of the rotating circle reaches the point at which it started (see figure 4.73).

Figure 4.73.

Since the coins are congruent, the question is, how often has the rolling coin rotated after it has traveled around the fixed coin exactly once? The typical response to this question is once. Yet this is a mistaken response.

What, then, is the correct answer to this question? We shall observe the movement of the center M_a of the moving circle at each of the quarter circles of the fixed circle. This can be seen in figures 4.74 through 4.78, where the point M_a travels along the dashed-line circle c_{aux} (*auxiliary*).

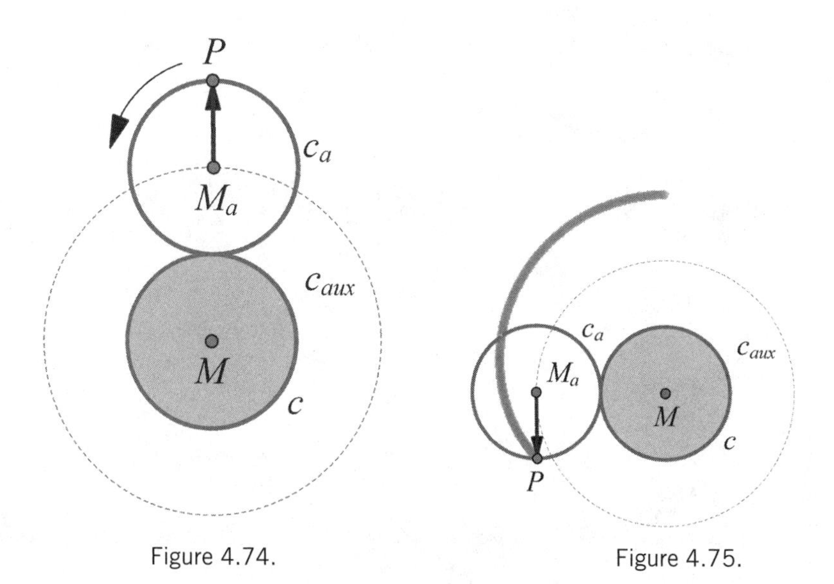

Figure 4.74. Figure 4.75.

In figure 4.76 we notice that the movable circle has made one rotation, since the point P is once again at the top. Consequently, when the rolling coin reaches its original position, it will have completed two revolutions. This is counterintuitive and consequently unexpected, which typically leads to a mistaken response to the original question. Incidentally, the path that the point P travels as called an epicycloid. Actually, since the two circles in question are the same size, this particular epicycloid, which has one cusp, is called a cardioid.

To provide a better picture of this movement, we provide increments of 90° rotations of the moving circle in figures 4.79 through 4.87.

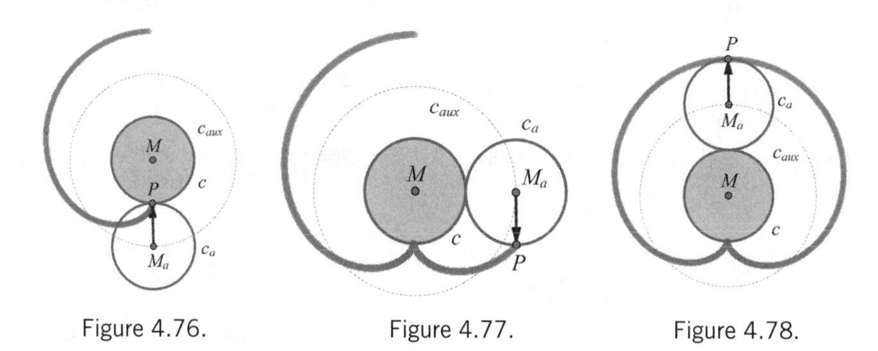

Figure 4.76. Figure 4.77. Figure 4.78.

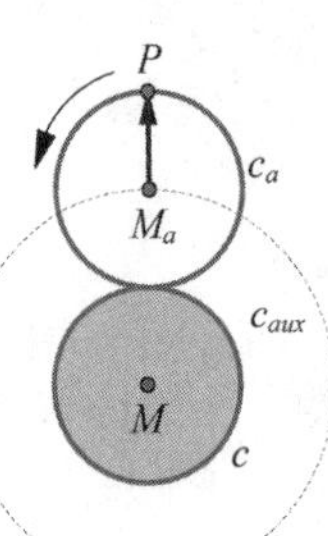

Figure 4.79.

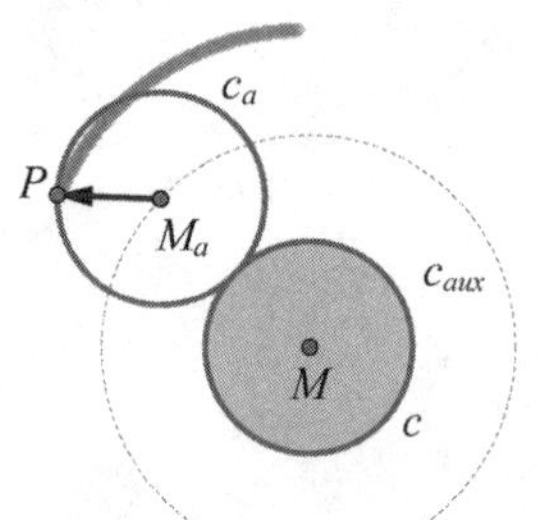

Figure 4.80.

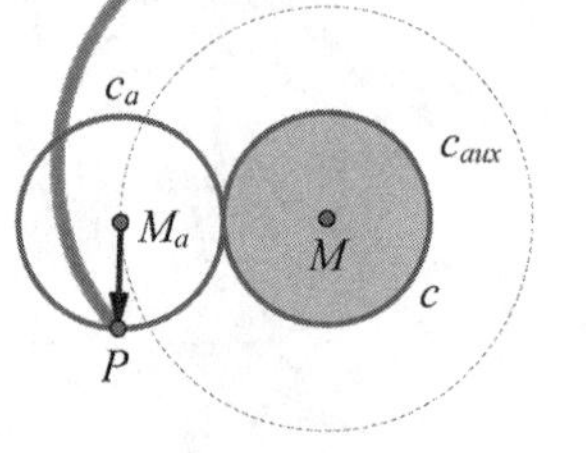

Figure 4.81.

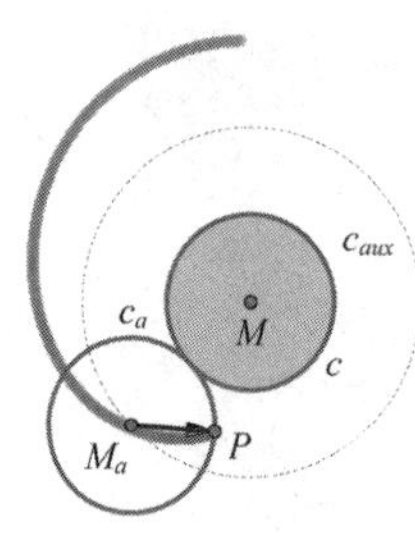

Figure 4.82.

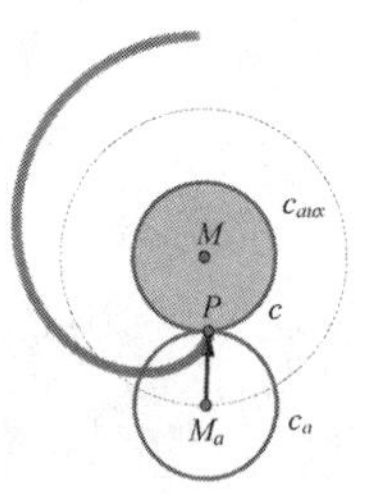

Figure 4.83.

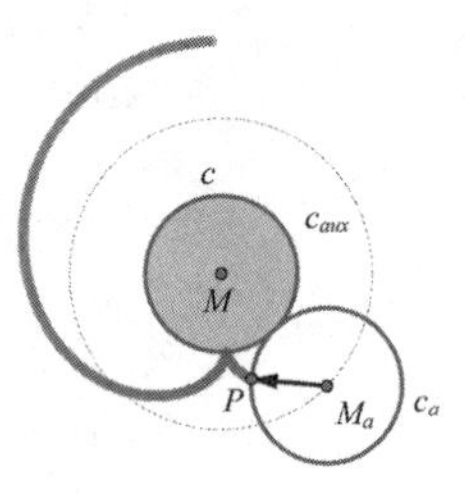

Figure 4.84.

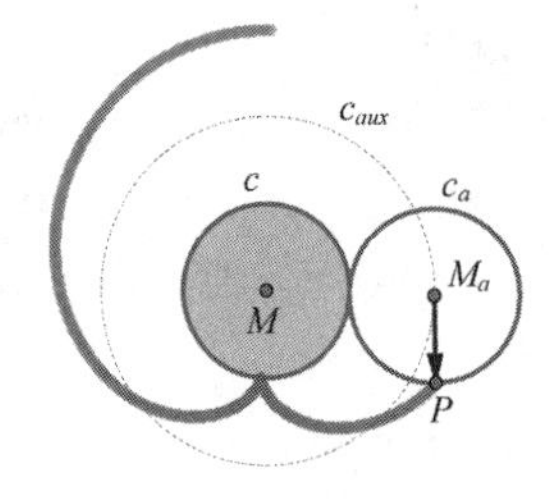

Figure 4.85.

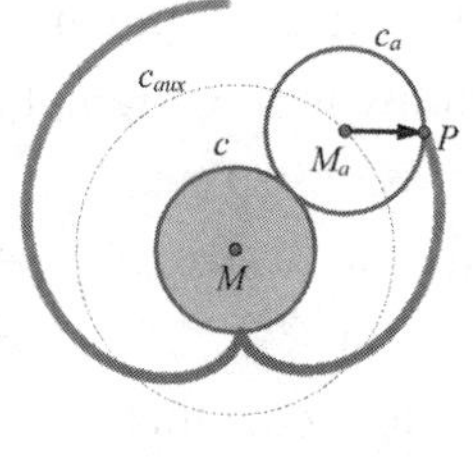

Figure 4.86.

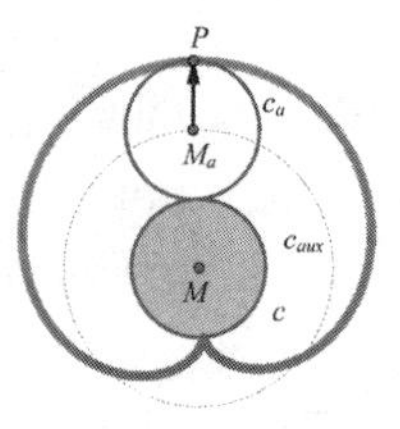

Figure 4.87.

You might want to try this by rotating one penny around another, keeping track all the while of the position of Abraham Lincoln's head as the moving penny rolls along—without slippage—about the fixed penny. This should help clear up the possible initial mistake.

A COMMON MISTAKE BASED ON A CORRECT PRINCIPLE

A basic concept in geometry is that the ratio of the areas of two similar figures is equal to the square of the ratio of two corresponding line segments. If we apply this principle to the following problem, we would run into a mistaken result. Let's consider that problem now.

We begin with two concentric circles whose radii are a and b, with $a > b$. Our task is to find the radius of a third concentric circle placed between the two given concentric circles such that the area of the outermost ring is twice the area of the next smaller ring (see figure 4.88). We will let the length of the radius of the sought-after circle be x.

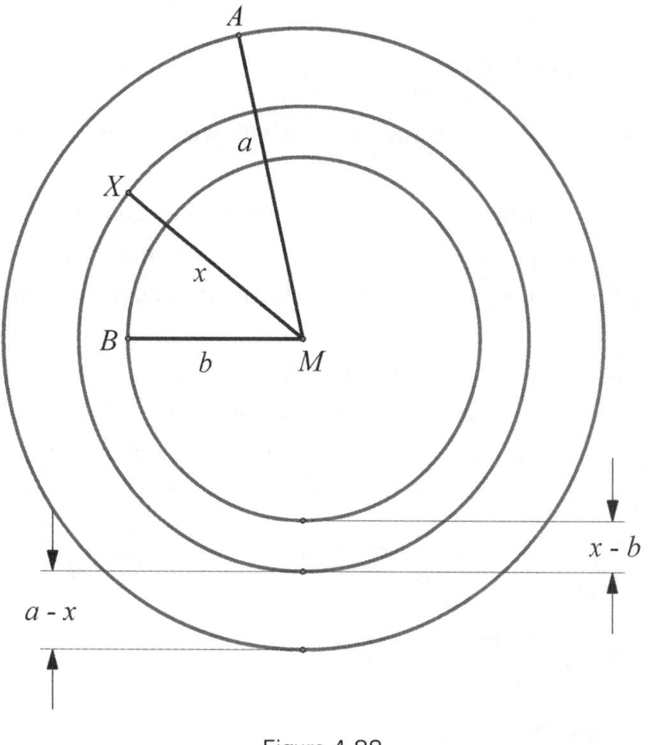

Figure 4.88.

Now applying the principle we mentioned at the start, the two corresponding parts could be the distances between the circles, or the width of the rings. Therefore, the outermost ring has a width of $a - x$, and the innermost ring has a width of $x - b$. We then get the following proportion:

$$\frac{(a-x)^2}{(x-b)^2} = \frac{2}{1}, \qquad (*)$$

Solving this equation for x, we get the following:

$$(a - x)^2 = 2 \cdot (x - b)^2$$
$$a^2 - 2ax + x^2 = 2 \cdot (x^2 - 2bx + b^2) = 2x^2 - 4bx + 2b^2$$
$$2x^2 - 4bx + 2b^2 - a^2 + 2ax - x^2 = 0$$
$$x^2 + 2x(a - 2b) - a^2 + 2b^2 = 0$$
$$x = -a + 2b \pm \sqrt{(-a+2b)^2 + a^2 - 2b^2} = -a + 2b \pm \sqrt{a^2 - 4ab + 4b^2 + a^2 - 2b^2}$$
$$x = -a + 2b \pm \sqrt{2a^2 - 4ab + 2b^2} = -a + 2b \pm \sqrt{2(a^2 - 2ab + b^2)}$$
$$x = -a + 2b \pm (a - b)\sqrt{2}.$$

Unfortunately, these two values of x are both wrong! Thus, we must have made a mistake somewhere.

Our mistake occurred at the very first step (marked with "*").

We mistakenly used the ring widths rather than the circles' radii to deal with the individual circles, whose areas would then be subtracted to get the ring areas.

If we let A_a, A_b, and A_x represent the areas of the circles whose radii are a, b, and x respectively, we can then set up the following with A_{a-x}, and A_{x-b} representing the areas of the two required rings as $A_{a-x} = A_a - A_x = \pi \cdot a^2 - \pi \cdot x^2 = \pi \cdot (a^2 - x^2)$, and $A_{x-b} = A_x - A_b = \pi \cdot x^2 - \pi \cdot b^2 = \pi \cdot (x^2 - b^2)$.

With $A_{a-x} = 2 \cdot A_{x-b}$, we get $\pi \cdot (a^2 - x^2) = 2\pi \cdot (x^2 - b^2)$, which then yields:

$$\frac{a^2 - x^2}{x^2 - b^2} = \frac{2}{1},$$

which is considerably different from the initial "application" of the similarity principle.

Now proceeding *correctly*, we get: $a^2 - x^2 = 2x^2 - 2b^2$, and therefore, $x^2 = \frac{a^2 + 2b^2}{3}$, which then gives us: $x = \sqrt{\frac{a^2 + 2b^2}{3}}$, since the negative root is ignored, as we are dealing with the length of a line.

A ROPE AROUND THE EQUATOR: A MISTAKE OF OUR INTUITION

A mistake in mathematics can also be one of judgment, where one makes a mistake because the correct answer is counterintuitive. Consider the planet Earth, with a rope wrapped tightly around the equator. Let's assume that Earth is a perfect sphere and that the equator is exactly 40,000 kilometers long. Assume also that Earth has a smooth surface along the equator—just to make our work a bit easier.

We now lengthen the rope by exactly one meter. We position this (now loose) rope around the equator so that it is uniformly spaced off Earth's surface (see figure 4.89). The question we are posed is, will a mouse fit under the rope?[7] It is quite expected that we *mistakenly* say that this is clearly not possible.

Figure 4.89.

The traditional way to determine the distance between the circumferences—of Earth and of the rope—is to find the difference between the radii. Let r be the length of the radius of the circle formed by Earth (circumference = C), and R the length of the radius of the circle formed by the rope (circumference = $C + 1$). See figure 4.90.

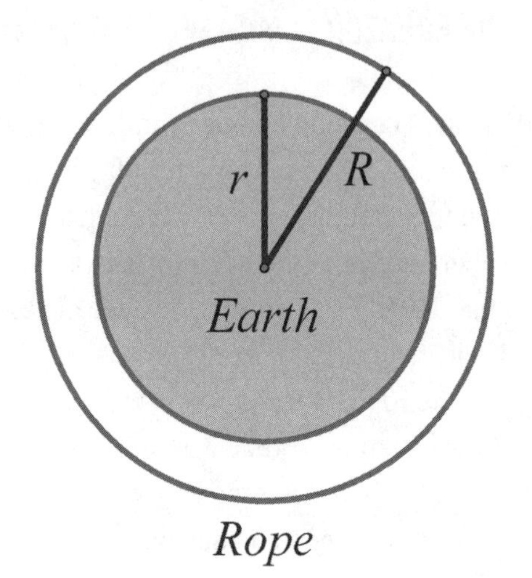

Figure 4.90.

The familiar circumference formulas give us:

$$C = 2\pi r, \text{ or } r = \frac{C}{2\pi}, \text{ and}$$

$$C + 1 = 2\pi R, \text{ or } R = \frac{C+1}{2\pi}.$$

We need to find the difference of the radii, which is:

$$R - r = \frac{C+1}{2\pi} - \frac{C}{2\pi} = \frac{1}{2\pi}.$$

The "1" in the numerator means one meter. Therefore, we get:

$$R - r = \frac{1\text{m}}{2\pi} = \frac{100\,\text{cm}}{2\pi} \approx 15.9 \text{ cm} = 0.159 \text{ m}.$$

Wow! There is actually a space of a bit over 6.25 inches for a mouse to crawl under.

You must really appreciate this astonishing result, as the intuitive answer from before would have clearly led to a mistake.

We might also have approached answering this question by using a very powerful problem-solving strategy that may be called *considering extreme cases*. You should realize that the solution was independent of the circumference or radius r of Earth, since the end result did not include the circumference in the calculation. It only required calculating $\frac{1}{2\pi}$.

Here is a really nifty solution using an extreme case. Suppose the inner circle (above) is very small, so small that it has a zero-length radius (which means it is actually just a point). We were required to find the difference between the radii; in this case, $R - r = R - 0 = R$. So, all we need to find is the length of the radius of the larger circle, and our problem will be solved. We apply the circumference formula for the circumference of the larger circle:

$$C + 1 = 0 + 1 = 2\pi R, \text{ then } R = \frac{1}{2\pi}.$$

This intially mistaken answer leads us to two lovely little treasures. First, it reveals an astonishing result—one clearly not anticipated at the start—and, second, it provides us with a nice problem-solving strategy that can serve as a useful model for future use.[8]

ANOTHER ROPE AROUND THE EQUATOR: FURTHER COUNTERINTUITION!

Up until now, the rope was always tightened concentrically, that is, it was pulled away evenly on all sides. Now this will not be the case. The rope will be pulled away at one single point, as if Earth were hung up on a hook.

The rope length is again extended by one meter. But again, instead of being concentric, it is pulled away at one point so a maximum distance from Earth's surface is achieved (see figure 4.91).

How far away from Earth can the rope be "pulled"?

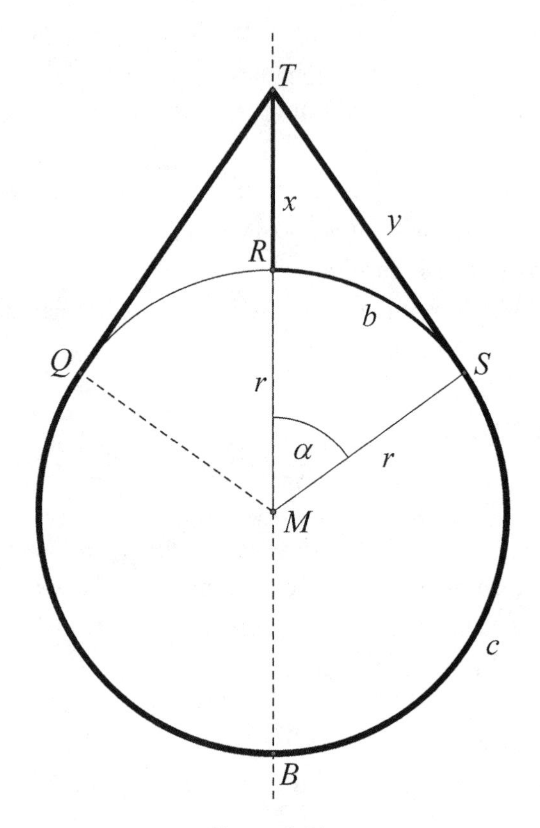

Figure 4.91.

Most people are astonished by the previous 15.9 centimeters of space for the mouse to crawl under. Now the result of pulling the rope from one point will once again be quite surprising. The result is apparently (once more) a paradox—one leading to mistaken conclusions.

The one-meter-longer rope pulled taut from a point, whereas the rest of the rope "hugs" Earth's surface, reaches a point about 122 meters above the earth's surface.

Let's see why this is so. This time the answer is clearly dependent on the size of Earth, and not exclusively on π—but remember π will also play a role in the issue. From the exterior point T, the rope (one meter longer than the circumference of the equator) is pulled taut so that it hugs Earth's surface until the points of tangency (S and Q). We seek to find how high off the surface the point T is. That means we will try to find the length of x or RT. Remember the length of the rope from B through S to T is 0.5 meter longer than the circumference of Earth. So that $BS + ST = BSR + 0.5$ m. We are going to try to find the length of TR.

So let's review where we are: The rope lies on the arc SBQ, which ends in the points S and Q and at points S and Q goes tangential to the point T. The lengths in the figure above are marked and $\alpha = \angle RMS = \angle RMQ$.

The length of the rope is $2\pi r + 1$, and we get the following relation: $y = b + 0.5$.

This is equivalent to $b = y - 0.5$ (y is 0.5 meter longer than b because of the extension by one meter).

In triangle MST, the tangent function will be applied as follows: $\tan \alpha = \frac{y}{r}$, so $y = r \cdot \tan \alpha$.

We can form the ratio of arc length to central angle measure and get the following:

$$\frac{b}{\alpha} = \frac{2\pi \cdot r}{360°}, \text{ and then we can get } b = \frac{2\pi \cdot r \cdot \alpha}{360°}$$

With $c = 2\pi r$ we can find Earth's radius (assuming that the equator is exactly 40,000,000 meters).

$$r = \frac{c}{2\pi} = \frac{40{,}000{,}000}{2\pi} \approx 6{,}366{,}198 \text{ [meters]}.$$

Combining the equations we have above, we get the following:

$$b = \frac{2\pi \cdot r \cdot \alpha}{360°} = y - 0.5 = r \cdot \tan \alpha - 0.5.$$

We are now faced with a dilemma, namely that this equation (obtained above) cannot be uniquely solved in the traditional manner. We will set up a table of possible trial values to see what will fit (i.e., satisfy the equation).

$\frac{2\pi \cdot r \cdot \alpha}{360°} = r \cdot \tan \alpha - 0.5$. We will use the value of r we found above, $r = 6{,}366{,}198$ meters.

α	$b = \dfrac{2\pi \cdot r \cdot \alpha}{360°}$	$b = r \cdot \tan \alpha - 0.5$	Comparison of Values (Number of Places in Agreement—**Bold**)
30°	**3**,333,333.478	**3**,675,525.629	1
10°	**1,1**11,111.159	**1,1**22,531.971	2
5°	**55**5,555.5796	**55**6,969.6547	2
1°	**111,1**11.1159	**111,1**21.8994	4
0.3°	**33,333**.33478	**33,333**.13940	5
0.4°	**44,444**.44637	**44,444**.66844	5
0.35°	**38,888.8**9057	**38,888.8**7430	6
0.355°	**39,444**.44615	**39,444**.45091	6
More Exactly:			
0.353°	**39,222.22**392	**39,222.22**019	7
0.354°	**39 333,335**04	**39,333.335**54	**8**
0.3545°	**39,388.89**059	**39,388.89**322	7
0.355°	**39,444**.44615	**39,444**.45091	6

Our various trials would indicate that our closest match of the two values occurs at $\alpha \approx 0.354°$.

For this value of α, $y = r \cdot \tan \alpha \approx 6{,}366{,}198 \cdot 0.006178544171 \approx 39{,}333.83554$ meters, or about 39,334 meters.

The rope is, therefore, almost forty kilometers long before it reaches its peak. But how high off Earth's surface is the rope? That is, what is the length of x?

Applying the Pythagorean theorem to triangle MST, we get $MT^2 = r^2 + y^2$.

$$MT^2 = 6{,}366{,}198^2 + 39{,}334^2$$
$$= 40{,}528{,}476{,}975{,}204 + 1{,}547{,}163{,}556 = 40{,}530{,}024{,}138{,}760.$$

Therefore, $MT \approx 6{,}366{,}319.512$ meters.

We are looking for x, which is $MT - r \approx 121.512$ meters, or about 122 meters.

This result is perhaps astonishing, because one intuitively assumes that by the circumference of the earth (40,000 kilometers) an extra meter must almost disappear. But this is the mistake! The larger the sphere, the farther the rope can be pulled away from it.

Looking at the extreme case, where the radius of the equator decreases to zero, we have the minimum value for x, namely $x = 0.5$ meters.

AN UNANTICIPATED MISTAKEN ASSUMPTION

One topic that has become part of the standard high-school geometry course is that of transformations: translations, rotations, and reflections. The notion of a reflection is usually done through a line. However, we can also reflect a figure in a circle. Typically, when one reflects a figure it remains the same type of figure. However, this can be a mistaken assumption when we reflect a triangle through a circle and expect to see its image as a triangle. We shall first review the rules for reflecting a point in a circle.

This reflection, also called an inversion, takes points and transforms them in both directions from inside and outside the circle. We will consider a circle with center M and radius r and a point outside the circle P, as shown in figure 4.92. To determine the reflection of the point P in the circle,we locate the point P' on ray MP such that $MP \cdot MP' = r^2$.

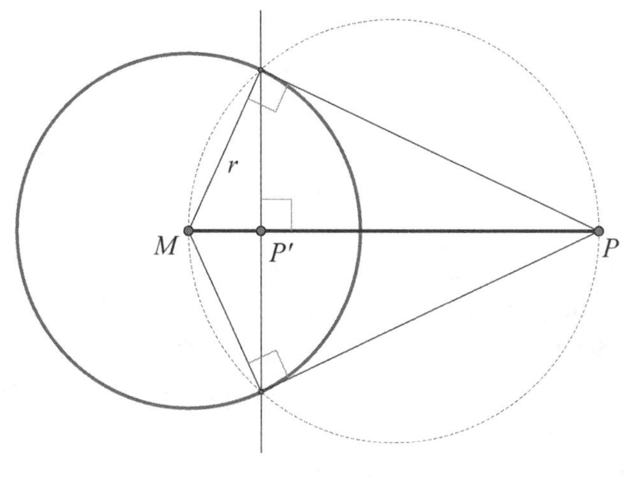

Figure 4.92.

If the point P is on the circle, then the reflection of P is the point itself. If the point P is inside the circle, then we construct a perpendicular to MP at P. At the points at which this perpendicular intersects the circle, we construct tangents to the circle, and the intersection of these tangents determines the reflection point P'.

On the other hand, if the point P is outside the circle, then we construct the two tangents to the circle from P. The chord joining the two points of tangency intersects the line MP at the reflection point P'. (See figure 4.92.)

We are now ready to consider the original problem, that of finding the reflection of a triangle in a circle. In figure 4.93, we show the reflections of the three vertices of triangle ABC in the circle as points A', B', and C', respectively. The temptation is to draw the triangle $A'B'C'$ and consider it the reflection of triangle ABC. Much to our surprise, this is a mistaken conclusion.

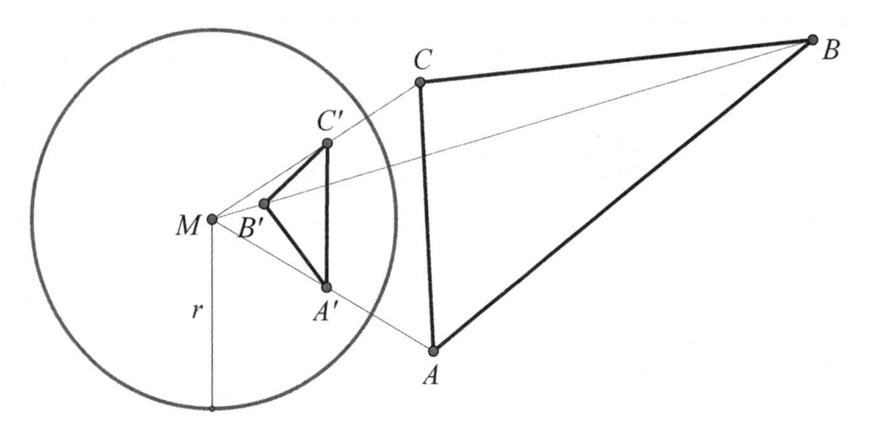

Figure 4.93.

If we now were to take the midpoints M_a, M_b, and M_c of the three sides of triangle ABC, we would get the points M'_a, M'_b, and M'_c, respectively. We notice that they do not lie on the side of the triangle $A'B'C'$, as shown in figure 4.94. This mistake should open up a desire for further study about this transformation.

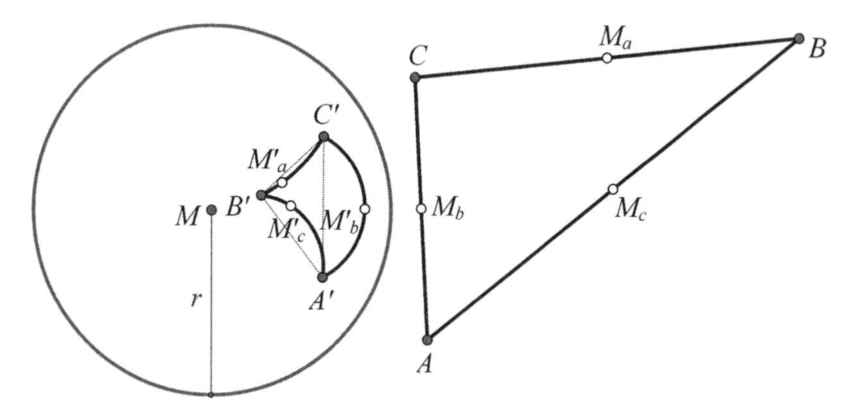

Figure 4.94.

The points on line through the center M of the circle of inversion (or reflection) will be reflected into a circle containing point M. That is why the reflection of segment AB results in arc $A'B'$ of the circle containing point M.

MISLEADING LIMITS

The concept of a limit is not to be taken lightly. It is a very sophisticated concept that can be easily misinterpreted. Sometimes the issues surrounding the concept are quite subtle. Misunderstanding of these can lead to some curious situations (or humorous ones, depending on your viewpoint). This can be nicely exhibited with the following two illustrations. Don't be too upset by the conclusion that you will be led to reach; remember, this is for entertainment. Consider them separately and then notice their connection.

It is simple to see that in figure 4.95, the sum of the lengths of the bold segments (the "stairs") is equal to $a + b$, since the sum of the vertical bold lines equals the length $OP = a$, and the sum of the horizontal bold lines equals $OQ = b$.

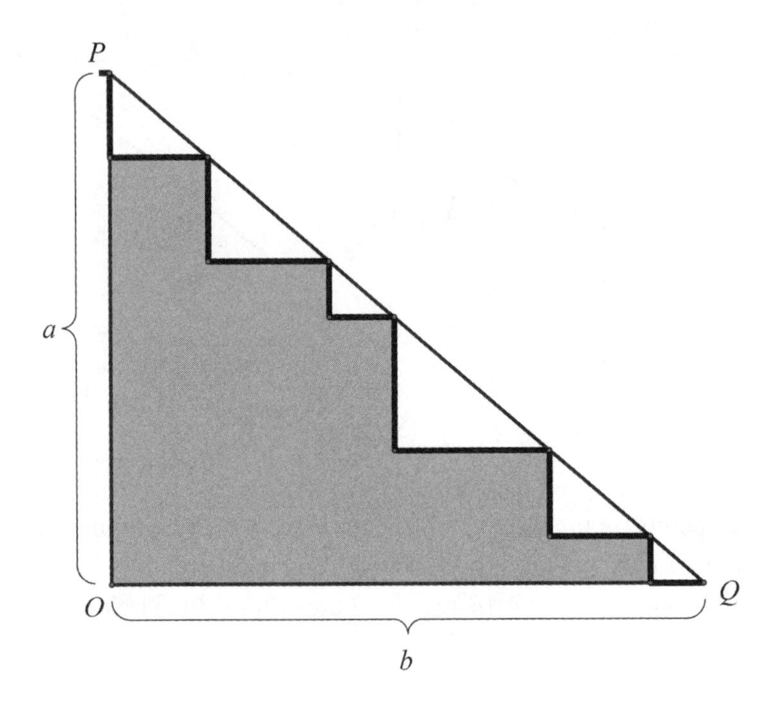

Figure 4.95.

The sum of the bold segments (“stairs”), found by summing all the horizontal and all the vertical segments, is $a + b$. If the number of stairs increases, the sum is still $a + b$. The dilemma arises when we continue to increase the stairs to a “limit” so that the stairs get smaller and smaller, which then makes the set of stairs appears to be straight line, in this case the hypotenuse PQ of triangle POQ. It would then appear that PQ has length $a + b$. Yet we know from the Pythagorean theorem that $PQ = \sqrt{a^2 + b^2}$ and *not* $a + b$. So what’s wrong?

Nothing is wrong! While the set consisting of the stairs does indeed approach closer and closer to the straight line segment PQ, it does *not* therefore follow that the *sum* of the bold (horizontal and vertical) lengths approaches the length of PQ, contrary to our intuition. There is no contradiction here, only a failure on the part of our intuition.

Another way to “explain” this dilemma is to argue the following. As the “stairs” get smaller, they increase in number. In the most extreme situation, we have stairs of 0 length in each dimension, used an infinite number of times, which then leads to considering $0 \cdot \infty$, which is meaningless! In truth, no matter how small the stairs get, the sum of two adjacent perpendiculars that form one of the small right triangles will never be equal to their hypotenuse. They will just be small right triangles. This may be a bit difficult to see, but that is one of the dangers of working with infinity.

Just as an aside, when considering the set of natural numbers, $\{1, 2, 3, 4, \ldots\}$, we would think that it is a larger set than the set of positive even numbers, $\{2, 4, 6, 8, \ldots\}$, because all the positive odd numbers are missing from the second set. Yet, since they are infinite sets, they are equal in size! We reason as follows: for every number in the set of natural numbers there is a member of the set of positive even numbers; hence they are equal in size. Counterintuitive? Yes, but that is what happens when we consider infinity.

It appears that infinity is playing games with us. The problem is that when we deal with infinity, we can no longer talk about the equality of sets the way we do when we have finite sets. The same is true with the staircase in our original problem. We can draw the finite steps, yet we cannot draw the infinite number of steps. Therein lies the problem.

A similar situation arises with the following example. In figure 4.96, the smaller semicircles extend from one end of the large semicircle's diameter to the other.

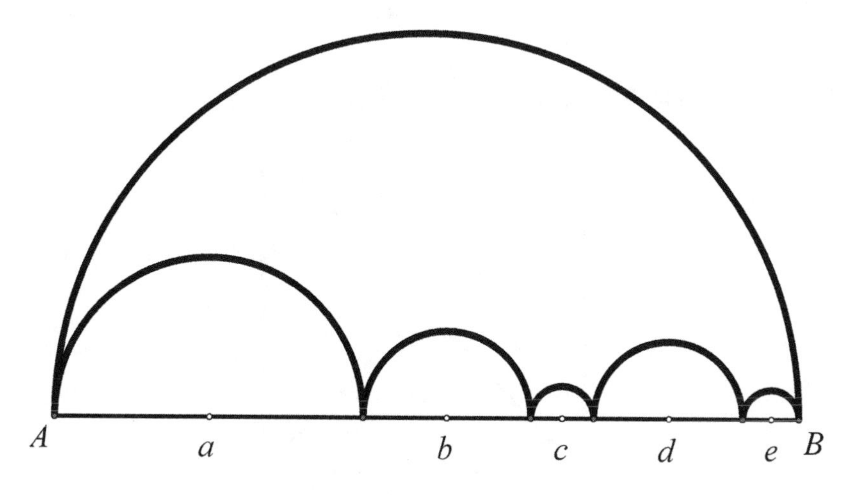

Figure 4.96.

It is easy to show that the sum of the arc-lengths of the smaller semicircles is equal to the arc-length of the larger semicircle. That is, the sum of the smaller semicircles

$$= \frac{\pi a}{2} + \frac{\pi b}{2} + \frac{\pi c}{2} + \frac{\pi d}{2} + \frac{\pi e}{2} = \frac{\pi}{2} \cdot (a + b + c + d + e) = \frac{\pi}{2} \cdot AB,$$

which is the arc length of the larger semicircle. This may not "appear" to be true, but it is! As a matter of fact, as we increase the number of smaller semicircles (where, of course, they get smaller) the sum "appears" to be approaching the length of the segment AB, that is, $\frac{\pi}{2} \cdot AB = AB$.

Taking this a step further, if we let $AB = 1$, then we have $\pi = 2$, which we surely know is a mistake!

Again, the set consisting of the semicircles does indeed appear to approach the length of the straight-line segment AB. It does *not* follow, however, that the *sum* of the semicircles approaches the *length* of the limit, in this case, AB.

This "apparent limit sum" is absurd, since the shortest distance between points A and B is the length of segment AB, not the semicircle arc AB (which equals the sum of the smaller semicircles). This is an important concept and may be best explained with the help of these motivating illustrations, so that future misinterpretations can be avoided.

WHICH ANSWER IS A MISTAKE?

Figure 4.97.

As we look at the diagram in figure 4.97, we ask which of the following potential descriptions of the figure is a mistaken answer?

1. A square with diagonals drawn.
2. The overhead view of a square right pyramid
3. The side view of a tetrahedron

None of these answers is a mistake, as you can see from the drawings in figures 4.98 through 4.100.

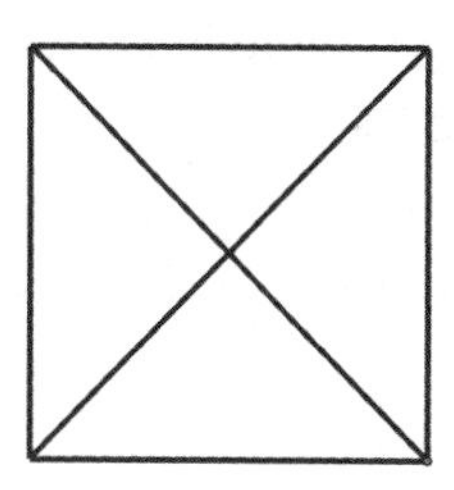

Figure 4.98.

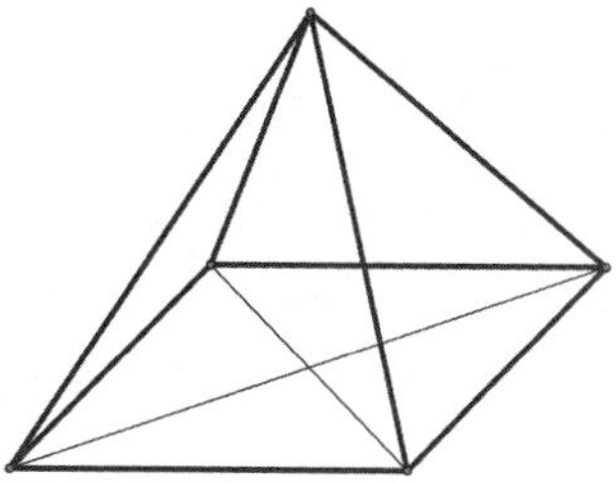

Figure 4.99.

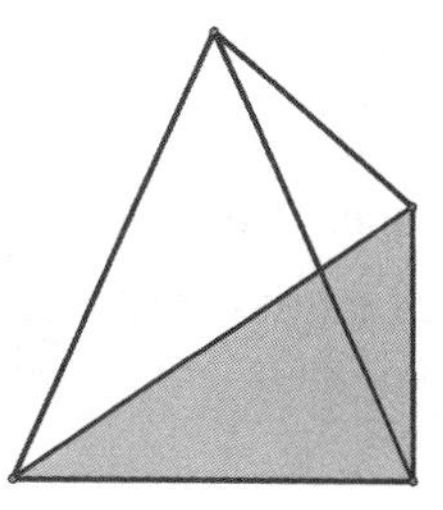

Figure 4.100.

A COMMON MISTAKE BASED ON INTUITION: COMBINING TWO PYRAMIDS

In pyramids *ABCD* and *EFGHI* shown in figures 4.101 and 4.102, respectively, all faces except base *FGHI* are equilateral triangles of equal size. If face *ABC* were placed on face *EFG* so that the vertices of the triangles coincide, how many exposed faces would the resulting solid have?

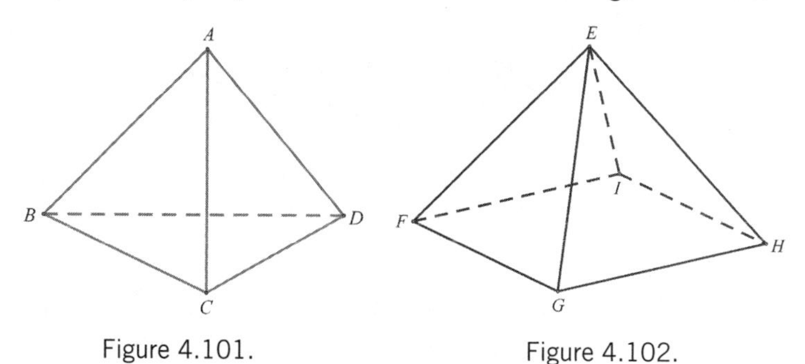

Figure 4.101.

Figure 4.102.

The typical way to approach this problem is to note that when we place the two pyramids together so that two of the faces coincide, and thereby are removed from the count of total number of faces, the expected answer would be 4 faces + 5 faces – 2 faces = 7 faces. This is a wrong answer, so there must be a mistake in our reasoning.

It should be noted that a national test included this question, including the wrong answer provided above, which had gone unnoticed for many years before was corrected.[9] It was not until a student discovered this error and persisted on his solution that this mistaken solution was brought to light. In fact, when the two solids are put together with their equilateral triangles overlapping, two rhombuses are formed, meaning that four equilateral triangle faces have now blended into two rhombus faces, as is shown in figure 4.103, where triangles *ACD* and *EFI* form one rhombus, and triangles *ABD* and *EGH* form a second rhombus. Therefore, the correct number of faces that results from the combining of the two solids is five.

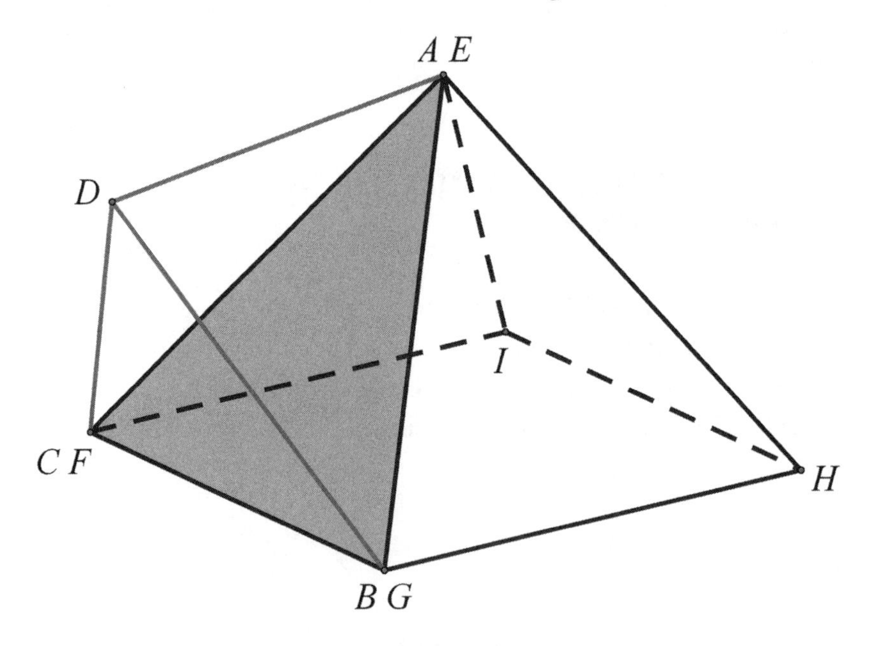

Figure 4.103.

The pyramid *EFGHI* with the square base *FGHI* and the equilateral triangle sides *EFG*, *EGH*, *EHI*, and *EFI* comprises one-half of a regular octahedron. Let us now consider another tetrahedron whose edges are twice as long as those of the above tetrahedron. If we cut off each of the four vertices by making cuts through the midpoints of each of the edges, we will have constructed a regular octahedron. We now take this octahedron and partition it into two pyramids with square bases. The five-faced solid, which came from one corner of the large tetrahedron and one-half of the middle part of the octahedron gives evidence that the conjectured rhombuses actually do exist on one plane and are not "bent" along a diagonal (see figures 4.104 and 4.105).

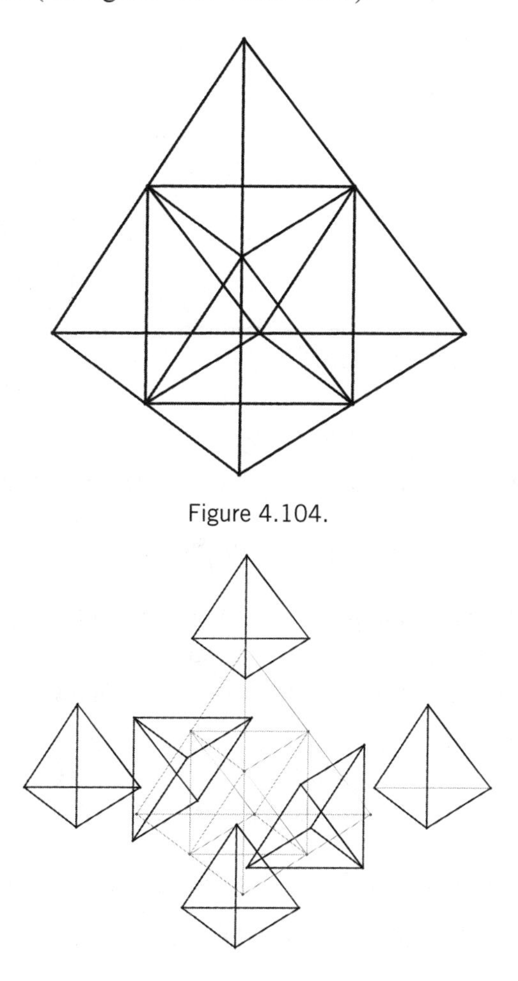

Figure 4.104.

Figure 4.105.

A STUDENT'S MISTAKE LEADS TO A CORRECT ANSWER[10]

A student is asked to cover with colored paper the inside and outside of an open box whose dimensions are $a = 20$ centimeters, $b = 10$ centimeters, and $c = 5$ centimeters, with the last measure being the height (figure 4.106). The question is, how much colored paper will be required to cover all these rectangular surfaces?

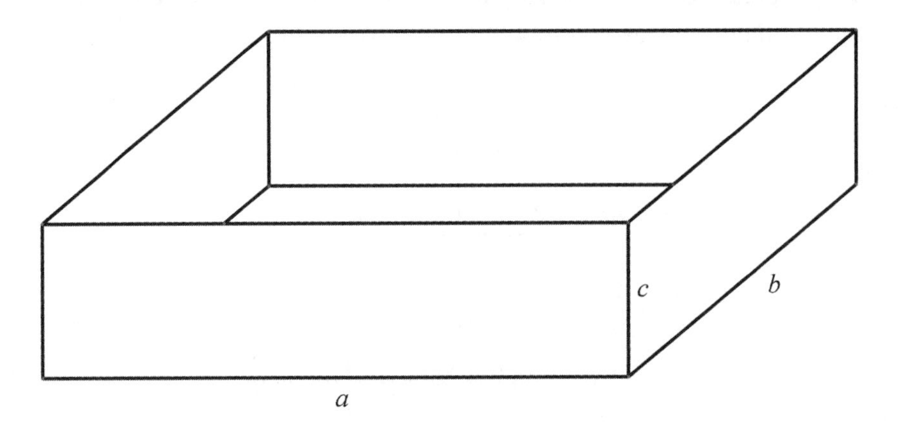

Figure 4.106.

Without much planning, the student does the following calculation:

$$A = a \cdot b \cdot c = 20 \cdot 10 \cdot 5 = 1{,}000 \text{ [cm}^2\text{]}.$$

Strangely enough, the answer is numerically correct! The question that then comes to mind is, how can this mistaken procedure lead to a correct answer? We know that the correct solution is as follows:

$$\begin{aligned} A &= 2(ab + 2ac + 2bc) = 2ab + 4ac + 4bc = 4c(a + b) + 2ab \\ &= 4 \cdot 5(20 + 10) + 2 \cdot 20 \cdot 10 = 1{,}000. \end{aligned}$$

Yes, we do get the same answer as the student. Will this always been the case? Under what circumstances will the following be true?

$$a \cdot b \cdot c = 4c(a + b) + 2ab$$

It turns out that there are fifty-six such triples of integers a, b, and c, where $a \geq b$. One such is $a = b = c = 10$ centimeters (an open cube), while another is $a = 220$ centimeters, $b = 5$ centimeters, and $c = 11$ centimeters.

OFTEN-MISTAKEN ATTEMPTS AT COMMON GEOMETRIC TRICKS

The question here is, what is the least number of straight lines required to connect the six points in figure 4.107 so that the lines are drawn without lifting your pencil off the paper?

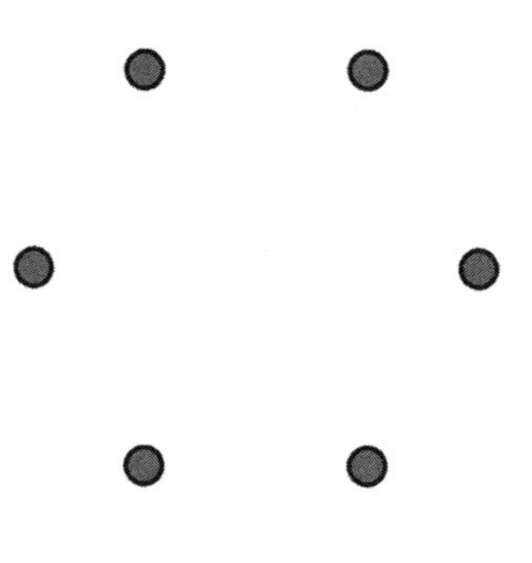

Figure 4.107.

A typical response to this question is five lines, usually drawn in one of the ways shown in figure 4.108. But is this the least number of straight lines that can be used to connect these six points?

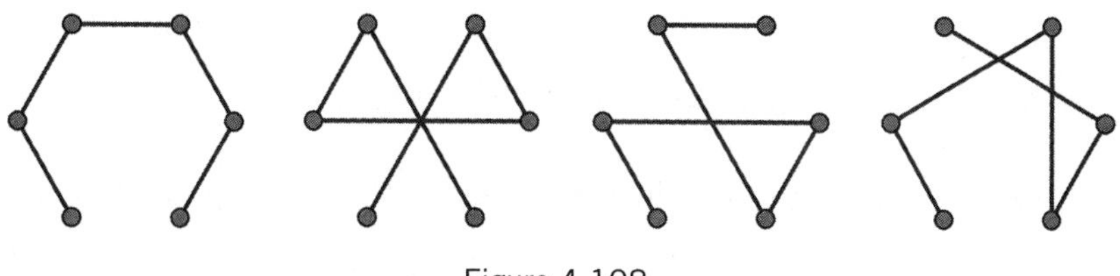

Figure 4.108.

As you might have expected, the answer is no. Fewer lines can be used to connect the six points.

The psychological mistake rests in the fact that we considered each line segments had to terminate at one of the points. As you can see from figure 4.109, we were able to connect the dots with four straight lines.

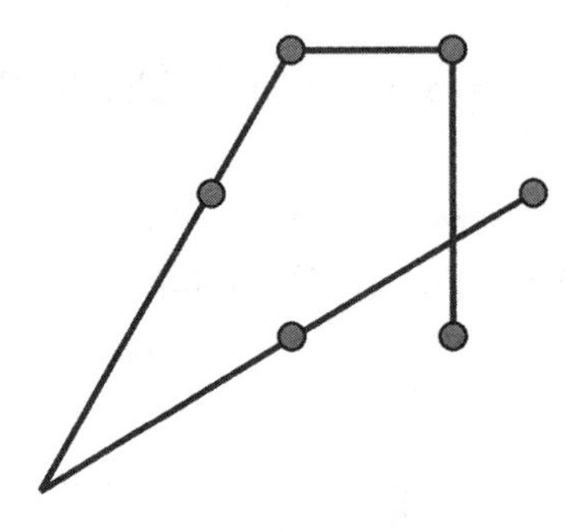

Figure 4.109.

Ridding us of the restriction of having line segments end at one of the given points even allows us to get a better solution, namely, using the three line segments we show in figure 4.110.

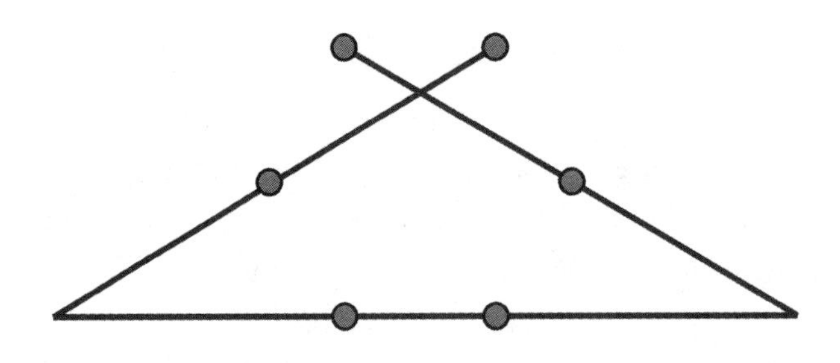

Figure 4.110.

Our earlier mistakes should now be instructive for the next situation.

This time, we are given nine dots, as shown in figure 4.111, and asked to connect them with four straight lines while not lifting the pencil off of the paper until the points are joined.

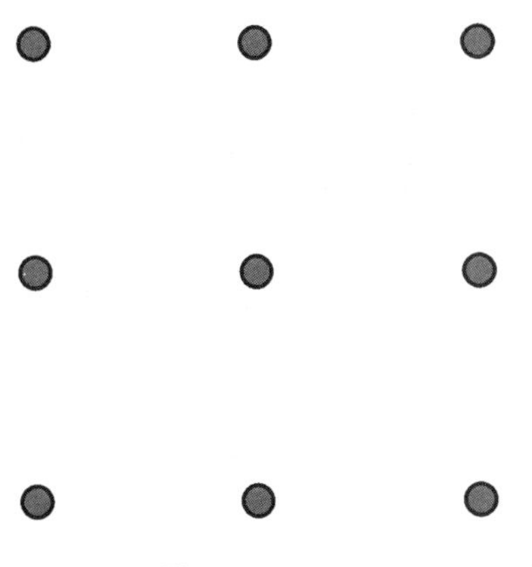

Figure 4.111.

Having learned from the earlier mistakes made above, we should be able to arrive at the solution offered in figure 4.112.

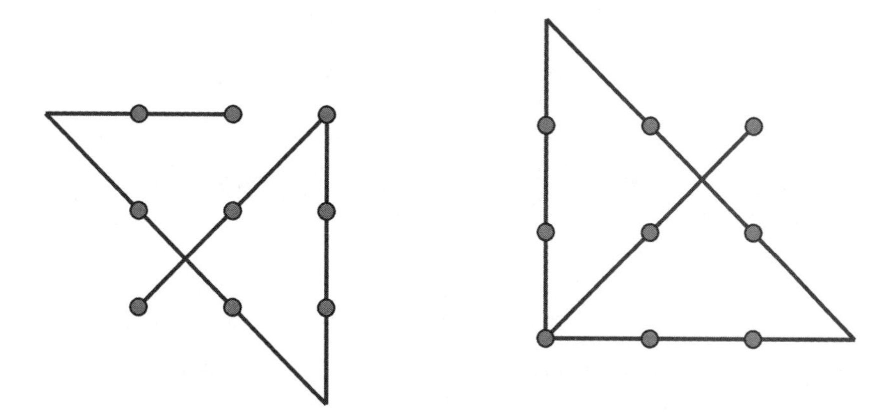

Figure 4.112.

Now that the reader will no longer make the mistake that most likely was made with the first of these dot-connecting problems, we offer two challenges.

1. Connect the twelve dots shown in figure 4.113 with as few as five straight lines, without lifting the pencil off of the paper, and return to the initial point.

Figure 4.113.

The solution is shown in figure 4.114.

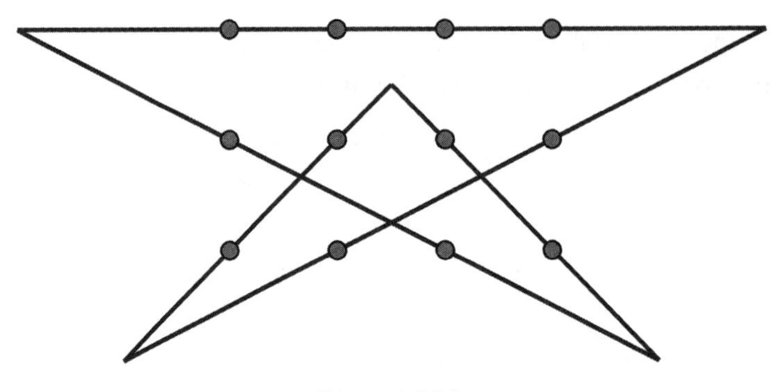

Figure 4.114.

2. Connect the twenty-five dots shown in figure 4.115 using only eight straight lines, without lifting the pencil off of the paper, and return to the initial point.

Figure 4.115.

Using nine lines would not be so difficult; however, with eight lines it becomes quite challenging. A solution is provided in figure 4.116.

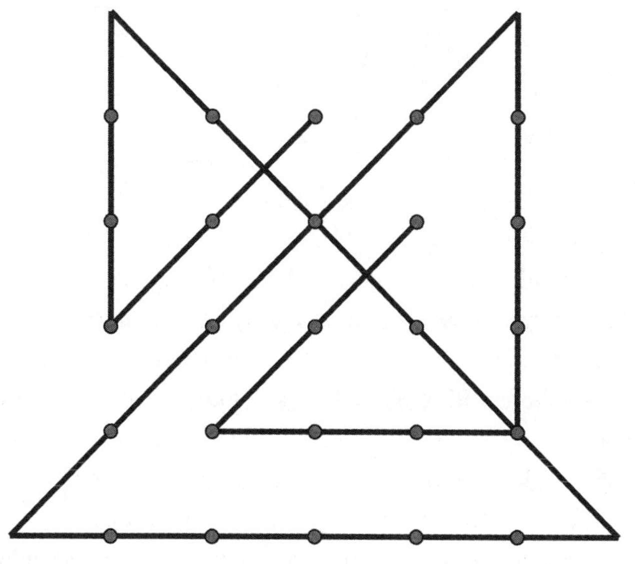

Figure 4.116.

AN EXPECTED MISTAKEN ANSWER

There is a question to which most people—including mathematicians—will give a mistaken answer. We are given the three shapes shown in figure 4.117, where we have a square with a side length 1, a circle with a diameter of length 1, and an isosceles triangle whose base and altitude have length 1. Is there a solid figure that, viewed from three different perspectives, has these three shapes? The typical, mistaken, answer is *no*. Yet you will see that there actually is such a solid figure.

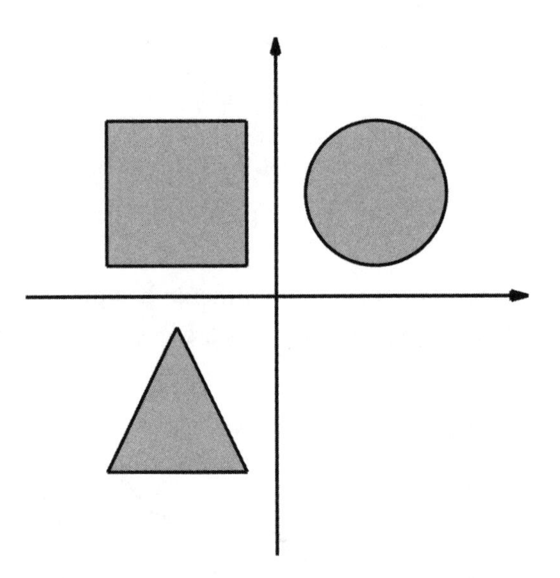

Figure 4.117.

To find a solid figure that has the shape of a square when viewed from three different directions is obviously easy to imagine. It is a cube with edge length 1, which will appear as a square when viewed from the three directions. A sphere with a diameter of 1 will appear as the circle shown in figure 4.117 when viewed from various directions. A cylinder with a height of length 1 and the base circle with diameter of 1 can also be seen from various directions as a square or as a circle, as shown in figure 4.117. Yet to get one solid that will render these three shapes when viewed from different directions seems nearly impossible and, therefore, usually produces a mistaken answer as we indicated.

As it is often expected, even teachers of mathematics find the challenge of finding one solid that can produce these three views nearly impossible, and thus offer a mistaken answer to the challenge.

Suppose we now cut these three shapes out of a board, as shown in figure 4.118. Let's see if we can find solid shapes that can be pushed through these figures with a tight fit.

Clearly, a cube can fit through the square as long as the edge length is appropriately sized. A sphere of diameter 1 or a cylinder of diameter 1 can easily be pushed through the circular cutout. A prism with a base congruent to the cutout of the isosceles triangle will also fit through the triangular cutout.

Figure 4.118.

However, the solid that we seek must be one that could be pushed through each of these three shapes with the expected tight fit. One possible solid is constructed by taking a cube with a unit-length edge, then cutting a cylinder out of it, as shown in figure 4.119. We then remove all the matter that is not included in a right prism.

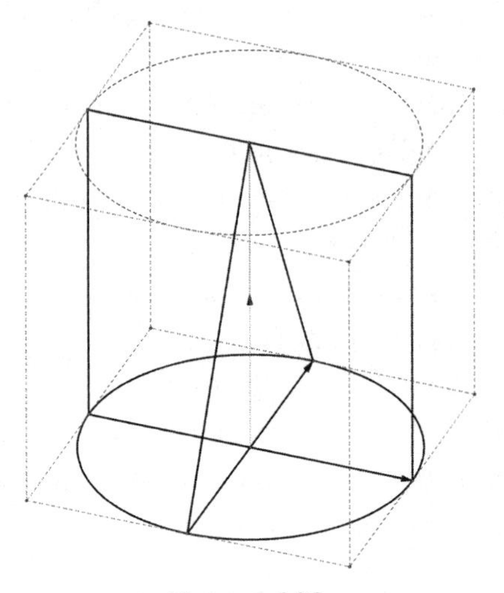

Figure 4.119.

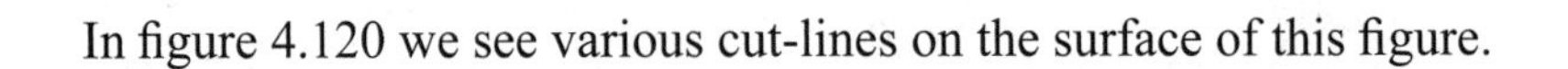

In figure 4.120 we see various cut-lines on the surface of this figure.

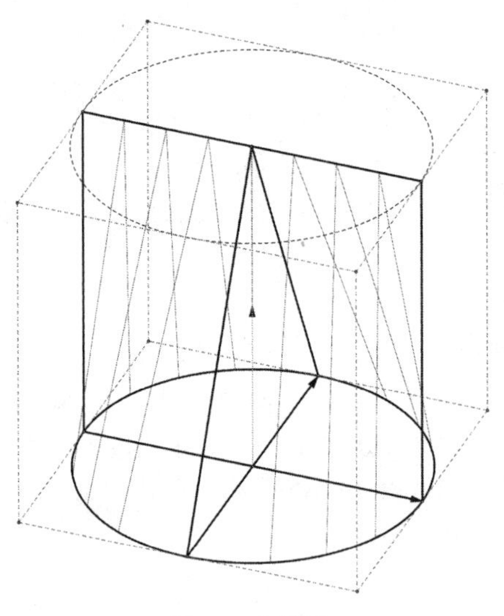

Figure 4.120.

In figures 4.121 to 4.124 we have a photograph of this resulting figure.

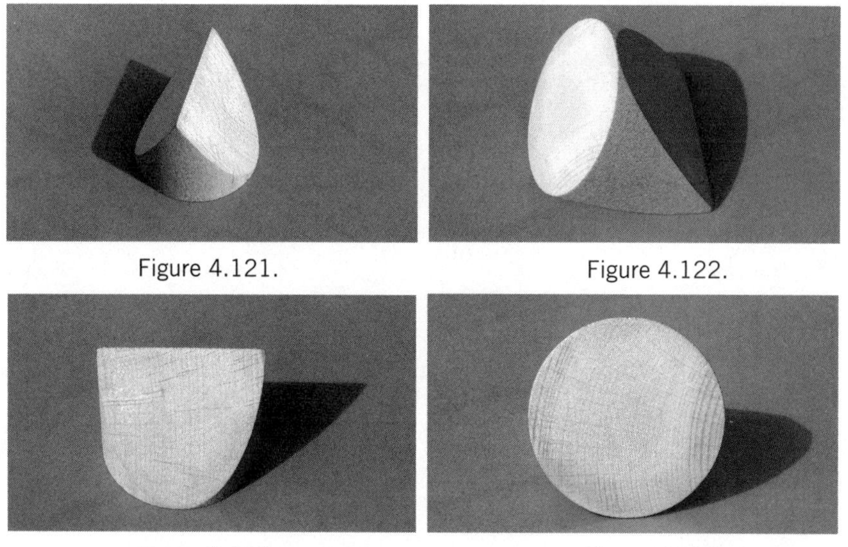

Figure 4.121.

Figure 4.122.

Figure 4.123.

Figure 4.124.

Figures 4.125, 4.126, and 4.127 show the required solid from the three directions that generate the three planar shapes we sought to create.

Figure 4.125.

Figure 4.126.

Figure 4.127.

As you can see, the viewpoint from which a solid figure is seen can be very important and possibly deceiving.

We have, thus, seen a wide variety of geometric mistakes. Many of these give us a much stronger view of geometric principles. Those that are seen as "paradoxes" also allow us to see the kind of misinterpretations often encountered without notice. In sum, through the journey of geometric mistakes our understanding and appreciation for geometry is hugely enhanced.

CHAPTER 5

MISTAKES IN PROBABILITY AND STATISTICS

In the course of our daily lives, we may often hear claims that one can "lie with statistics." This usually refers to an argument that emphasizes statistics favorable to one's point and downplays those that are unfavorable. Mark Twain has popularized a now often-used quote in his "Chapters from My Autobiography" (published in 1906 in the *North American Review*) when he wrote that "There are three kinds of lies: lies, damned lies, and statistics." We can see that this branch of mathematics has gotten a bad rap. In this chapter, we will present a number of misconceptions that seem to support this claim because they are conclusions based on mistaken thinking or mistaken computation. In any case, the mistakes we present will at first upset, but then (hopefully) will satisfy an understanding of the proper processes of probability and statistical thinking.

A QUESTION THAT CAN LEAD TO MISTAKEN CONCLUSIONS: DO BOYS HAVE MORE SISTERS THAN GIRLS HAVE?

We tend to reason by example. Suppose we consider a family of two children, a boy and a girl. The boy has a sister, while the girl does not. In this case, the boy has more sisters than the girl has. Now let's take a family with three children, a boy and two girls. The boy has two sisters, while each of the sisters has only one sister. This kind of reasoning leads us to the conclusion that boys have more sisters than girls have. However, this reasoning is mistaken.

To see where correct reasoning leads to the conclusion that boys and girls have the same number of sisters, consider a family with two children. Here the possibilities are to have two boys, two girls, or a boy and a girl. In the last case, we would have to consider them twice, since it could be a boy and a girl, or a girl and a boy—order being important. Another way of saying this is that there are four cases—each with equal probability: boy-boy, boy-girl, girl-boy, and girl-girl. In the first case, that of two boys, the boys have zero sisters. In the case of two girls, and where there are no boys, we once again have zero sisters for boys. In the case of a boy and a girl, the boy has one sister—times two, because again, there are two cases to consider. Therefore, in the family of two children, there are two boys' sisters, and two sisters' sisters—the latter coming from the family of two girls. Therefore, for a family of two children, there would appear to be an equal number of sisters for boys and sisters for girls.

To further make the case, we can consider a family of three children. We summarize this in the table below.

Combination	Number of Combinations	Boys' Sisters	Girls' Sisters
boy, boy, boy	1	0	0
boy, boy, girl	3	$2 \cdot 1$	0
boy, girl, girl	3	2	$2 \cdot 1$
girl, girl, girl	1	0	$3 \cdot 2$

These three-children families could have twelve boys' sisters (see the second and third rows of the table), and twelve girls' sisters (see the third and fourth rows of the table). Once again, we have the same number of sisters for boys and girls. We can generalize that to a family of n children, they will contribute $n(n-1)2^{n-2}$ sisters for both boys and girls.

This question was clearly not as simple as it appeared to be at the outset and was ripe for a mistake.

MISTAKES IN COMBINATORICS AND PROBABILITY

Here we present two problems that easily lead to some unanticipated mistakes. They can serve as an alert to what will be coming in this chapter.

Problem I: From six players how many different three-player team games can be formed? This is a question in combinatorics, and we will look at two different solutions and see if the mistaken one can be discovered before we expose it.

Solution 1:

To form the first team of three from six players, we get the following directly from simple combination rules:

$$ {}_6C_3 = \binom{6}{3} = \frac{6 \cdot 5 \cdot 4}{1 \cdot 2 \cdot 3} = 20, $$

which tells us that there are twenty possible teams that can be formed. The second team is automatically formed after the first has been chosen by simply using the remaining three players. Therefore, the answer is simply that there are twenty ways in which the three players can be chosen from the six players.

Solution 2:

If the six players are represented by A, B, C, D, E, and F, we can show the various teams by the listing them as we have in the table below.

Team 1	Team 2		Team 1	Team 2
A, B, C	D, E, F		A, C, E	B, D, F
A, B, D	C, E, F		A, C, F	B, D, E
A, B, E	C, D, F		A, D, E	B, C, F
A, B, F	C, D, E		A, D, F	B, C, E
A, C, D	B, E, F		A, E, F	B, C, D

This table shows that there are ten different teams that can be formed. This seems to be a significant difference between the results of solution 1 and solution 2. Thus, there must be a mistake somewhere.

It turns out that solution 1 is wrong, exhibiting a very important error in combinatorics. When we said in solution 1 that we chose three players and the remaining three players select themselves by default, we did not take into account the fact that the first three selected are each duplicated once throughout the twenty variations. That is, if ABC is one team, then DEF is the second team. However, when DEF is selected, the other team is ABC by default. Thus, the teams ABC/DEF appear twice in the calculation of solution 1. Therefore, solution 2 is the correct one.

Problem II: During the production of one hundred electric switches, it was found that five of them were faulty. It turns out that of these one hundred switches, three have already been sold. What is the probability that these three sold switches are all defective?

Solution 1:

To select three switches from the one hundred switches given, we use the combination rules to get:

$${}_{100}C_3 = \binom{100}{3} = \frac{100 \cdot 99 \cdot 98}{1 \cdot 2 \cdot 3} = 161{,}700,$$

which is the number of ways in which three switches from the one hundred can be selected.

The number of ways that we can select the three switches from the five defective ones is:

$${}_5C_3 = \binom{5}{3} = \frac{5 \cdot 4 \cdot 3}{1 \cdot 2 \cdot 3} = 10.$$

Therefore, the probability that all three selected switches are defective is $\frac{10}{161{,}700} = \frac{1}{16{,}170} \approx 0.00006$.

Solution 2:

We will begin by letting P(S) represent the probability of selecting the good switches, and P($\overline{S}$) represent the probability of selecting the defective switches. We can then establish the probability of selecting each of these kinds of switches as follows:

$$P(S) = \frac{95}{100} = \frac{19}{20} = 0.95 \text{ and } P(\overline{S}) = \frac{5}{100} = \frac{1}{20} = 0.05.$$

Then the probability of selecting three defective switches in a row is the product of those three probabilities:

$$P(\overline{S}) \cdot P(\overline{S}) \cdot P(\overline{S}) = \frac{1}{20} \cdot \frac{1}{20} \cdot \frac{1}{20} = \frac{1}{8{,}000} = 0.000125.$$

We now have two different solutions to the original question. The second one is clearly a mistake.

Individually, each of the probabilities P(S) and P($\overline{S}$) is correct when it comes to selecting only the first switch. However, if the first switch we sold was a defective switch, we remain with ninety-nine switches, of which four are defective switches. Therefore, the probability that the second switch being sold is also defective is not $\frac{5}{100}$; rather it is $\frac{4}{99}$. It then follows that the probability that the third switch being sold is defective is $\frac{3}{98}$. Therefore, the probability of three consecutive defective switches being sold is the product of the three individual probabilities:

$$P(\overline{S_1}) \cdot P(\overline{S_2}) \cdot P(\overline{S_3}) = \frac{5}{100} \cdot \frac{4}{99} \cdot \frac{3}{98} = \frac{1}{16{,}170} = 0.00006.$$

When we compare this answer to solution 1, we clearly see that solution 2 was a mistake, one that is frequently made when determining such probabilities.

MISTAKEN REASONING IN A MINI VERSION OF SUDOKU

In the game of sudoku, the object is to place the numerals in such a way that no numeral is repeated in any three-by-three box, row, or column. In our abbreviated model for this example, we will be using smaller boxes and only the numerals 1, 2, 3, and 4.

Our task is to place the numerals 1–4 in the squares shown in figure 5.1 so that no numeral is repeated in any row or column, nor in any of the quadrants I, II, III, and IV. The question we need to respond to here is, in how many ways can this task be completed?

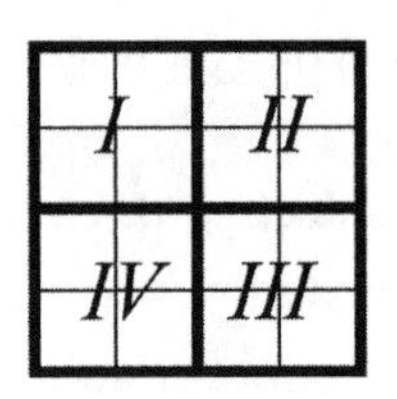

Figure 5.1.

Solution 1:

We shall begin by filling in the squares in quadrants I and III (see figure 5.2).

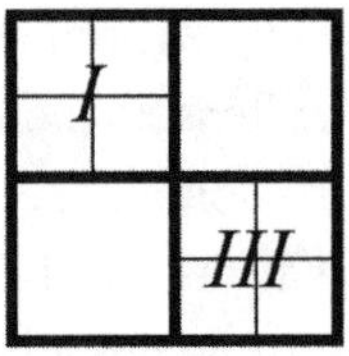

Figure 5.2.

Filling quadrant I can be done in 4! = 24 ways. We can complete the insertion of numbers into quadrant III also in 4! = 24 ways. Completing these two quadrants I and III are two independent events. Therefore, we can use the product rule, and we find that there are $24 \cdot 24 = 576$ possibilities to complete these two quadrants. The remaining two quadrants would be filled in for each of these 576 possibilities so as to avoid having any numeral appear twice in any row, column, or quadrant. As an example of how this would work, consider figure 5.3 as one of the 576 possibilities and how it leads to figure 5.4. Similarly, figure 5.5 would lead to figure 5.6.

1	2		
3	4		
		1	2
		3	4

Figure 5.3.

1	2	4	3
3	4	2	1
4	3	1	2
2	1	3	4

Figure 5.4.

4	2		
3	1		
		1	4
		2	3

Figure 5.5.

4	2	3	1
3	1	4	2
2	3	1	4
1	4	2	3

Figure 5.6.

In this fashion we have established that there are 576 possibilities to complete this game.

Solution 2:

Our next solution method is to complete placement of the numerals 1–4 in quadrant I as shown in figure 5.7.

1	2		
3	4		

Figure 5.7.

By considering all possibilities, starting with this formation in quadrant I, we get twelve possibilities, as shown in figure 5.8.

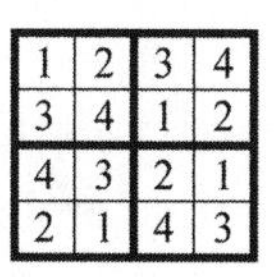

1	2	4	3
3	4	2	1
4	3	1	2
2	1	3	4

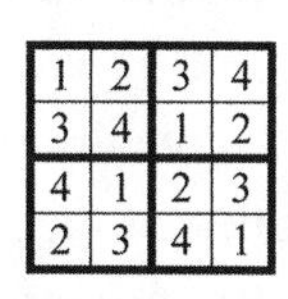

1	2	3	4
3	4	2	1
4	3	1	2
2	1	4	3

1	2	4	3
3	4	2	1
2	3	1	4
4	1	3	2

1	2	4	3
3	4	1	2
4	3	2	1
2	1	3	4

1	2	3	4
3	4	1	2
4	3	2	1
2	1	4	3

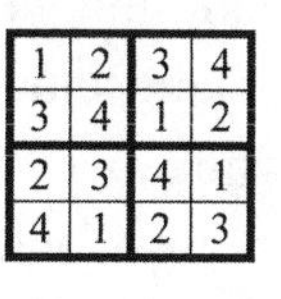

1	2	3	4
3	4	1	2
4	1	2	3
2	3	4	1

1	2	4	3
3	4	2	1
4	1	3	2
2	3	1	4

1	2	4	3
3	4	2	1
2	1	3	4
4	3	1	2

1	2	4	3
3	4	1	2
2	1	3	4
4	3	2	1

1	2	3	4
3	4	1	2
2	3	4	1
4	1	2	3

1	2	3	4
3	4	2	1
2	1	4	3
4	3	1	2

1	2	3	4
3	4	1	2
2	1	4	3
4	3	2	1

Figure 5.8.

Since we know that the number of possibilities for placing the numerals in quadrant I is $4! = 24$, we can use the product rule to show that the number of possible placements of numerals in this four-by-four square is $24 \cdot 12 = 288$. In other words, there are 288 possible squares that would allow us to successfully complete the game.

We now have two answers to our original question. Which of these answers is right, and which one has a mistake in its solution? Well, it turns out that solution 1 is mistaken. We are just unable to fill the remaining quadrants II and IV by holding to the requirement of the lack of repetition of numerals. In figure 5.9, we would be unable to complete quadrant IV. We have made a false assumption in solution 1.

1	2		
3	4		
		1	3
		4	2

Figure 5.9.

Any half-filled puzzle leads to exactly one solution, or to none. This can be seen in that solution 2, which is correct, has exactly half of the number of valid solutions as are found in the mistaken solution 1. These are considerations that must be taken to avoid obvious mistakes.

THE FAMOUS BIRTHDAY PROBLEM: AN INTUITIVE MISTAKE

Here we present one of the most surprising results in mathematics. It is one of the best ways to convince the uninitiated of the "power" of probability. Aside from being entertaining, examining this problem may upset your sense of intuition and help you avoid making anticipated mistakes.

Let us suppose you are in a room with about thirty-five people. What do you think the chances (or probability) are of at least two of these people having the same birth date (month and day only)?

Intuitively, one usually begins to think about the likelihood of two people having the same date out of a selection of 365 days (assuming no leap year). Translating into mathematical language: 2 out of 365 would be a probability of $\frac{2}{365} = .005479 \approx \frac{1}{2}$ percent. A minuscule chance. Is this correct or a mistake?

Let's consider a randomly selected birthday group: the first thirty-five presidents of the United States. We chose that number of people since it can represent the size of a rather large class of students. You may be astonished that there are two presidents with the same birth date: The eleventh president, James K. Polk (November 2, 1795), and the twenty-ninth president, Warren G. Harding (November 2, 1865), shared the same birthday.

Figure 5.10. James K. Polk. Figure 5.11. Warren G. Harding.

You may be surprised to learn that for a group of thirty-five, the probability that at least two members will have the same birth date is greater than eight out of ten, or $\frac{8}{10} = 80$ percent. In other words, our intuition would have led us to a mistaken conclusion. Before we see how this can be calculated, let's see if it really holds true.

If you have the opportunity, you may wish to try your own experiment by selecting ten groups of about thirty-five people each to check on birthday matches. You ought to find that in about eight of these ten groups there will be a match of birthdates. For groups of thirty people, the probability that there will be a match is greater than seven out of ten; or in seven of these ten groups there would be a match of birth dates. What causes this incredible and unanticipated result? Can this be true? It seems to go against our intuition.

How can this probability be so high when there are 365 possible birthdates? (Actually there are 366 possible birthdates, but for the sake of our illustration—and simplicity—we will not address the possibility of birthdays on February 29.) Let us consider the situation in detail and walk through the reasoning that will eventually convince us that these are the true probabilities. Consider a class of thirty-five students. What do you think is the probability that one selected student matches his own birth date? Clearly *certainty*, or 1.

This can be written as $\frac{365}{365}$.

The probability that another student does *not* match the first student (i.e., has a different birth date) is $\frac{365-1}{365}=\frac{364}{365}$.

The probability that a third student does *not* match the first and second students is $\frac{365-2}{365}=\frac{363}{365}$.

The probability of all thirty-five students *not* having the same birth date is the *product* of these probabilities:

$$p=\frac{365}{365}\cdot\frac{365-1}{365}\cdot\frac{365-2}{365}\cdot\ldots\cdot\frac{365-34}{365}.$$

Since the probability (q) that at least two students in the group *have* the same birth date and the probability (p) that *no* two students in the group have the same birth date is a certainty (i.e., there is no other possibility), the sum of those probabilities must be 1, which represents certainty.

Thus, $p + q = 1$, and it follows that $q = 1 - p$.

In this case, by substituting for p, we get:

$$q=1-\left(\frac{365}{365}\cdot\frac{365-1}{365}\cdot\frac{365-2}{365}\cdot\ldots\cdot\frac{365-34}{365}\right)\approx 0.8143832388747152.$$

In other words, the probability that there will be a birth date match in a randomly selected group of thirty-five people is somewhat greater than $\frac{8}{10}$. This is, at first glance, quite unexpected when one considers there were 365 dates from which to choose. The motivated reader may want to investigate the nature of the probability function. The table below provides a few values to serve as a guide:

Number of People in Group	Probability of a Birth-Date Match	Probability (in Percent) of a Birth-Date Match
10	.1169481777110776	11.69%
15	.2529013197636863	25.29%
20	.4114383835805799	41.14%
25	.5686997039694639	56.87%
30	.7063162427192686	70.63%
35	.8143832388747152	81.44%
40	.891231809817949	89.12%
45	.9409758994657749	94.10%
50	.9703735795779884	97.04%
55	.9862622888164461	98.63%
60	.994122660865348	99.41%
65	.9976831073124921	99.77%
70	.9991595759651571	99.92%

Notice how quickly almost-certainty is reached. With about sixty students in a room, the chart indicates that it is almost certain (99 percent) that two students will have the same birth date.

Were one to do this with the death dates of the first thirty-five presidents, one would notice that two died on March 8 (Millard Fillmore in 1874 and William H. Taft in 1930) and that three presidents died on July 4 (John Adams and Thomas Jefferson in 1826, and James Monroe in 1831). This latter coincidence could be interpreted by some that a death date could be willed.

From the table above, we see that in a group of thirty people, the probability of there being two people with the same birth date is about 70.63 percent. Yet, if we change this to the situation where you go into a room with thirty people and look for someone in the room with the same birthday as you, then we can determine the probability to be about 7.9 percent—considerably lower, since we now seek a specific birth date, rather than just a match of any birth date.

Let's see how this can be found. We will determine the probability that there are no matches to your birth date and then subtract that probability from 1.

$$p_{\text{Probability that no one has your birthday}} = \left(\frac{364}{365}\right)^{30}$$

The probability that none of these thirty people has your birth date is then:

$$q = 1 - p_{\text{Probability that no one has your birthday}} = 1 - \left(\frac{364}{365}\right)^{30} \approx 0.0790085980895507\overline{69}.$$

Perhaps even more amazing is that if you have a randomly selected group of two hundred people in a room, the probability of having two of these people born on the very same day (i.e., same year as well!) is about 50 percent.

Above all, this astonishing demonstration should serve as an eye-opener about the inadvisability of relying too much on intuition and thereby help you to avoid making a magnificent mistake!

THE SOCCER-PENALTY-SHOT DILEMMA: AVOIDING A MISTAKE!

A soccer team's goalkeeper in two consecutive soccer matches was able to establish the following achievements: in the first game, he blocked two of five penalty shots; in the second game, he blocked two of three penalty shots. We are being asked to determine his success rate in blocking penalty shots.

We will now present three possible solutions to this question, two which are mistaken solutions and only one of which is correct.

Solution 1:

We can look at the problem by combining his performance in the two matches and thereby as his having blocked four of eight penalty-shot attempts. This would allow us to get the following result:

$$\frac{2}{5} \oplus \frac{2}{3} = \frac{2+2}{5+3} = \frac{4}{8} = 0.5.$$

Solution 2:

We can also look at his success rate in each of the two games and then add them to get:

$$\frac{2}{5} + \frac{2}{3} = \frac{16}{15} \ (\approx 1.07).$$

However, $\frac{16}{15} > 1$, which would indicate that the goalkeeper is better than 100 percent in his blocking of penalty shots. This is obviously a mistake that should be immediately recognized.

Solution 3:

This time we will try to equalize the two soccer matches. This can be done by equating the number of penalty shots at fifteen (this being the lowest common denominator of the two fractions), and then proportionally increasing the number of blocked shots. That would indicate that in the first match, he would have blocked six of fifteen penalty shots; and in the second match, he would have blocked ten of fifteen penalty shots. This would then lead to the following: $\frac{6}{15} \oplus \frac{10}{15} = \frac{6+10}{15+15} = \frac{16}{30} = \frac{8}{15}$ (≈ 0.53).

We are now faced with the question of determining which is the correct solution.

Solution 1 is clearly correct, while solution 2 is certainly mistaken. Solution 3 could give us some information, but solution 1 is clearly the best way of determining the answer to the question.

WHAT IS THE RIGHT PATH TO B-A-S-K-E-T-B-A-L-L?

From the distribution of letters in figure 5.12 we are being asked to determine how many paths will spell out the word *basketball*. We are expected to do this by beginning at the top with the letter *B*, and ending at the bottom of the letter *L*.

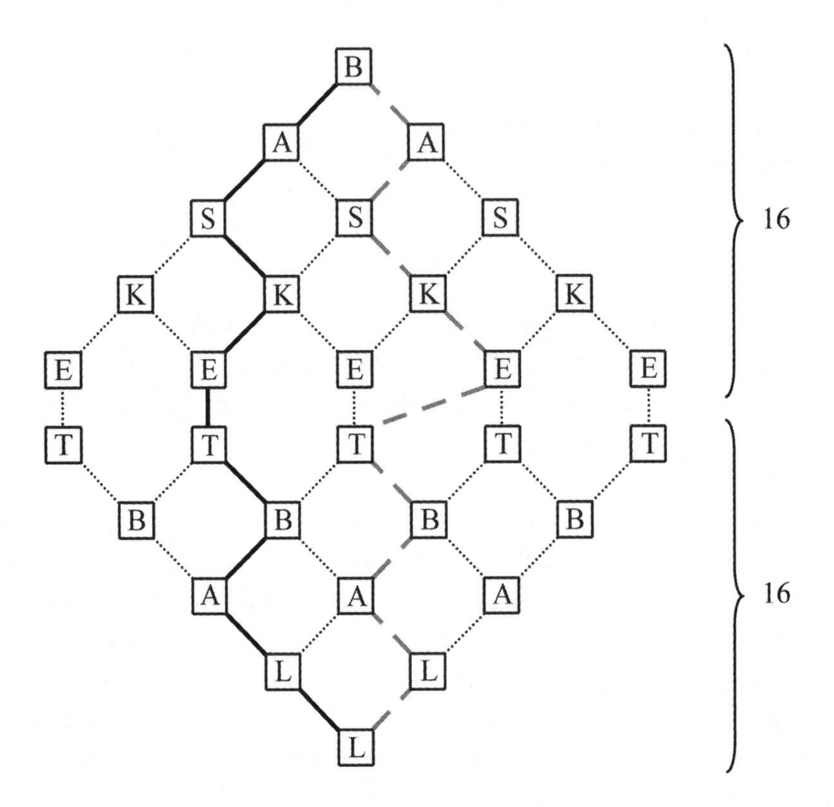

Figure 5.12.

Solution 1:

With a systematic set of attempts, we will find that we can have a number of paths that are:

2 BA, 4 BAS, 8 BASK, and 16 BASKE (see figure 5.12). If we were to start from the bottom and work our way upward from the *L*, we would find that we have 16 LLABT paths, as shown in figure 5.12. We might then conclude that we have sixteen paths in one direction going toward the center, and 16 paths in the opposite direction also going toward the center, and, therefore, we would have $16 \cdot 16 = 256$ complete paths.

Solution 2:

We could consider each letter as an end position and then inspect how many paths there are from the *B* to that particular letter. We show these path numbers next to each letter in figure 5.13.

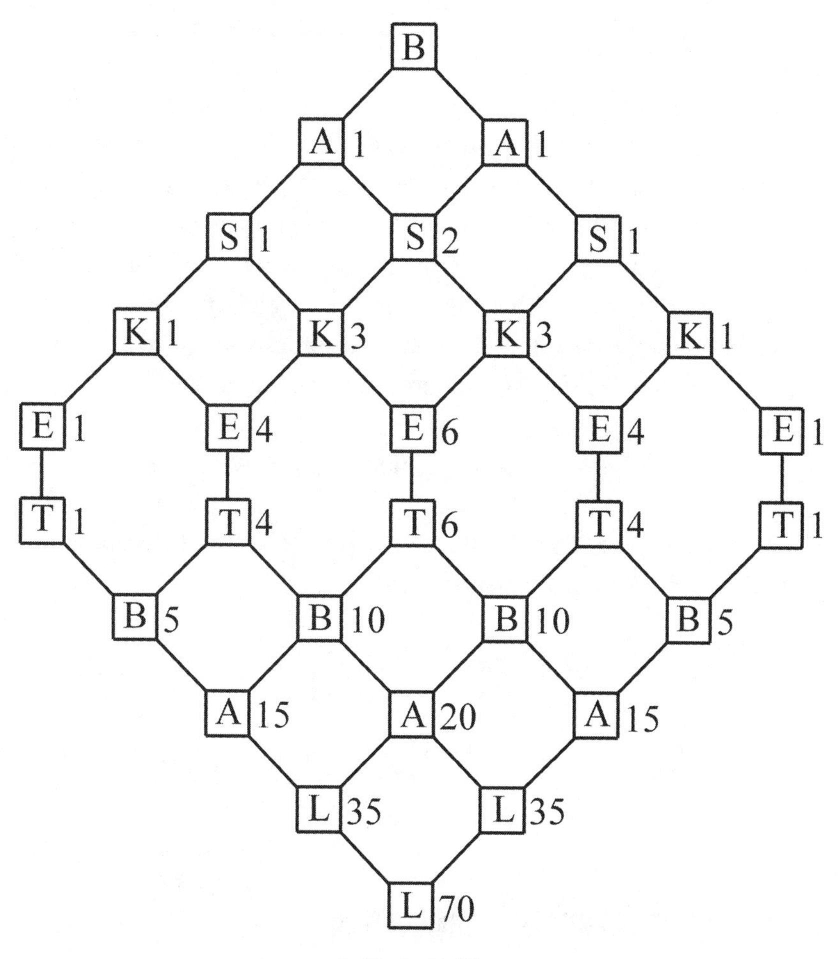

Figure 5.13.

This can be easily determined because of the symmetry of the figure. In other words, each subsequent number of paths can be determined by adding the two above the letter to the right and left of it. For example,

$2 = 1 + 1$, $3 = 1 + 2$, $4 = 1 + 3$, $6 = 3 + 3$. Completing this chart in this fashion as we show in figure 5.13, we can complete the chart and find that the word *basketball* can then be created by seventy different paths.

Once again we are faced with a question, which of these solutions is correct and which is mistaken?

If you said the first solution is wrong, then you were right. The conclusion that we can draw by merely assuming that every path of BASKE from the top, and every path LLABT from the bottom can be uniquely connected is false. Although this would be possible, as shown with the bold markings in figure 5.12, it must not necessarily be the case as shown in this figure where the dashed connecting lines show an incorrect connection between the *E* and the *T*. The latter paths were incorrectly counted in our calculation $16 \cdot 16 = 256$. That is the mistake in solution 1.

We could correct the first solution by dealing with the two five-letter portions separately, as shown in figure 5.13. We would then get $1 \cdot 1 + 4 \cdot 4 + 6 \cdot 6 + 4 \cdot 4 + 1 \cdot 1 = 70$.

The second solution leads to a correct answer. We can easily support the rule that we used here. Consider figure 5.14. All the paths that lead to the *Z* must go through *X* or *Y*. If there are *x* paths that lead to *X*, and *y* paths that lead to *Y*, then the paths that lead to *Z* is: $z = x + y$.

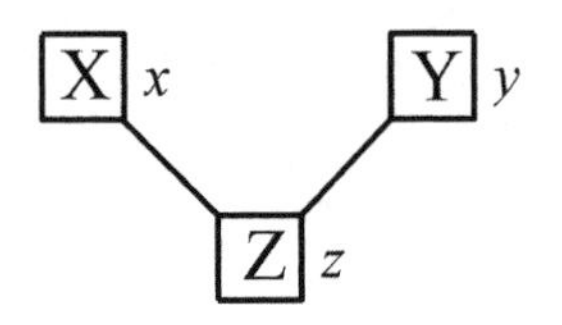

Figure 5.14.

THE MISTAKE OF IMPRECISE FORMULATION

We begin with an equilateral triangle inscribed in a circle. If a chord of the circle is selected at random, what is the probability that the chord length is greater than the side of the equilateral triangle? (See figure 5.15.) This question was first posed by the French mathematician Joseph Bertrand

(1822–1900). He gave three possible answers, all apparently correct, yet yielding different results.

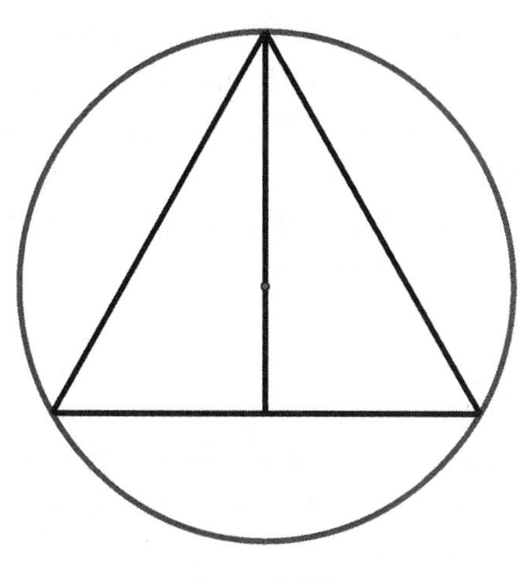

Figure 5.15.

Method 1:

We begin this method by choosing two random endpoints on the circle and then drawing a chord joining them (figure 5.16). We then draw an equilateral triangle inscribed in the circle and having one vertex at one of the endpoints of the chord as shown in figure 5.17.

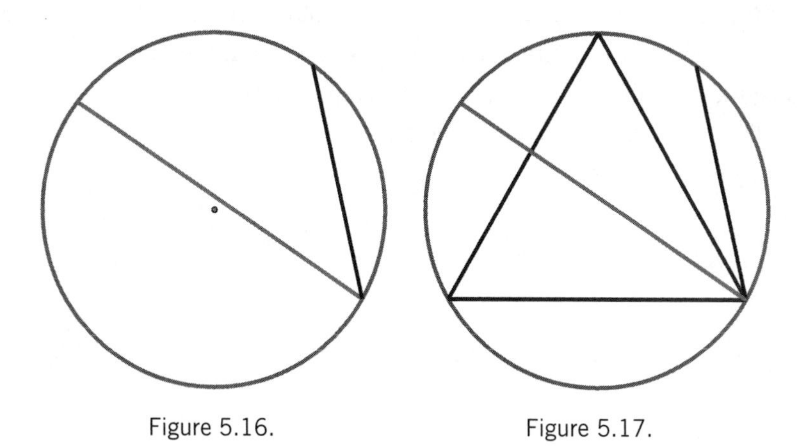

Figure 5.16. Figure 5.17.

If the chord originally drawn has its other endpoint on the arc of the circle formed by the opposite side of the equilateral triangle, then the chord is longer than one of the sides of the equilateral triangle (see figure 5.17). On the other hand, if the randomly drawn chord has the other endpoint on one of the arcs of the circle formed by the adjacent sides of the equilateral triangle, then the chord will be shorter than one of the sides of the triangle. Since the arc of the circle that will create a longer chord is $\frac{1}{3}$ of the circumference of the circle, the probability that the randomly drawn chord is longer than the side of the triangle is equal to $\frac{1}{3}$.

Method 2:

This method for determining whether a randomly drawn chord in a circle is shorter or longer than a side of the inscribed equilateral triangle begins with drawing a diameter *AB* of the circle. We then select a point on this diameter (C_1 or C_2, . . .) and draw a chord perpendicular to the diameter *AB* at this point (see figure 5.18). We then draw an inscribed equilateral triangle so that one side is perpendicular to this diameter. As a matter of fact, the side of the equilateral triangle perpendicular to the diameter also bisects the radius *MA* (see figure 5.19). Therefore, the chord perpendicular to the radius and outside the equilateral triangle will be shorter than the side of the equilateral triangle, and a chord that is perpendicular to the radius and cuts the other two sides of the equilateral triangle will be longer than the side of the equilateral triangle. In other words, half the time the randomly drawn cord could be on either side of the midpoint of the radius. Thus, the probability is $\frac{1}{2}$ that the randomly drawn chord will be longer than the side of the equilateral triangle.

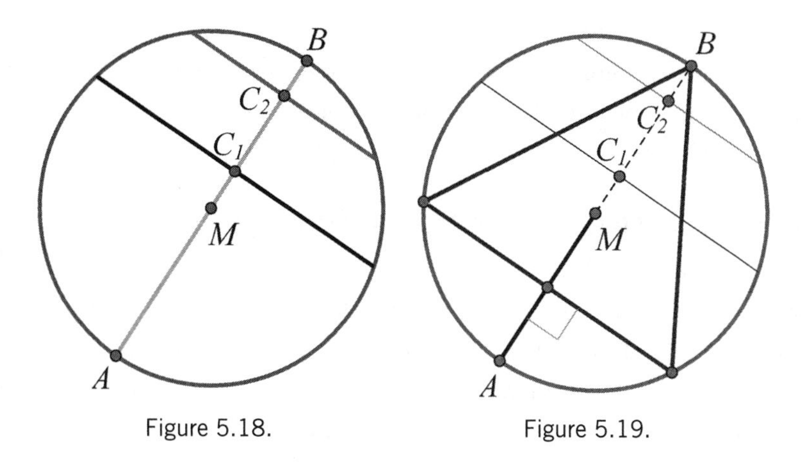

Figure 5.18. Figure 5.19.

Method 3:

Here we will select any point within the circle and then construct a chord with this chosen point as the midpoint (see figure 5.20). Our next step is to draw an equilateral triangle inscribed in the circle, and then inscribe a circle in the equilateral triangle (see figure 5.21). It turns out that the radius of the inscribed circle is $\frac{1}{2}$ the length of the radius of the circumscribed circle. Therefore, the area of the inscribed circle is $\frac{1}{4}$ of the area of the circumscribed circle of the equilateral triangle. Thus, the original chord selected is longer than the side of the equilateral triangle, if the chosen midpoint is in the inscribed circle. Therefore, the probability of the point falling inside the inscribed circle is $\frac{1}{4}$.

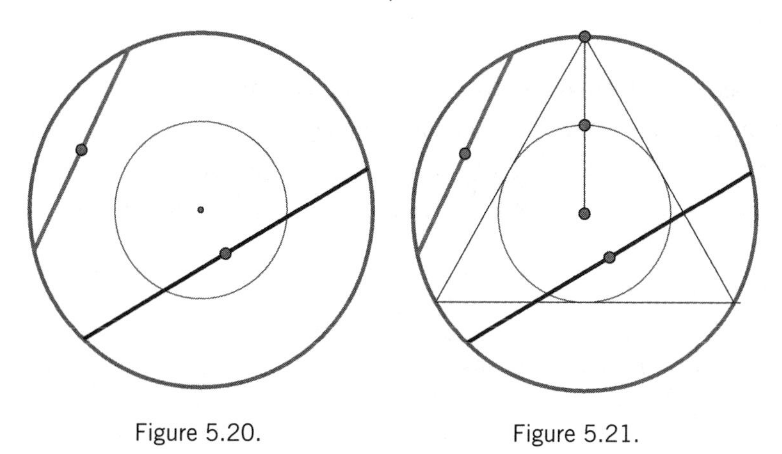

Figure 5.20. Figure 5.21.

This paradox actually has no answer because the randomness of the original selection of the chord was not well defined. Therefore, astonishingly, the mistake here is one of a lack of a proper definition.

THE UNMISTAKEN CONCLUSION OF A BETTING GAME

Two gamblers are playing a game in which they each have an equal chance of winning. The first player to win five games is considered the champion. At any point in this multigame match, player A has won four games and player B has won three games. Players agreed to end the match at this point. How should the champion money be divided between the two players? Let's look for possible mistakes.

Solution 1:

In this solution we will consider the possibilities for the remaining games, were they to have been played. The winners could be A, BA, BB. In the first two cases, A will have won the championship; while in the last case, B would have been the champion. Thus, the ratio of their winning likelihoods is 2 : 1. And then follows that A should get $\frac{2}{3}$ of the champion money, and B should get $\frac{1}{3}$ of the champion money.

Solution 2:

Since each of the players has the same probability of winning each game, the distribution of the championship money should be in the ratio of the current winnings, namely, 4 : 3. It therefore follows that the money should be distributed with A getting $\frac{4}{7}$ and B getting $\frac{3}{7}$ of the championship money.

Solution 3:

In order for B to win the championship, two more games would have to be played. This can occur in the following ways indicating the winners: BB,

BA, *AB*, *AA*. In the first case, *B* would be the champion; while in the last three cases, *A* would be the champion. This would give us a ratio of 3 : 1. Therefore, the money would be split in the following way: *A* would get $\frac{3}{4}$ of the championship money, and *B* would get $\frac{1}{4}$ of the championship money.

You will notice of the three solutions, we come up with three different answers. Only one of them can be right, yet intuitively they all make sense. Which one is correct and which are mistakes?

If we figure the probabilities of the three cases in solution 1, we get:

$$\mathrm{P}(A) = \frac{1}{2},\ \mathrm{P}(BA) = \frac{1}{2} \cdot \frac{1}{2} = \frac{1}{4},\ \text{and}\ \mathrm{P}(BB) = \frac{1}{2} \cdot \frac{1}{2} = \frac{1}{4}.$$

We can see that the three cases are not equally probable. More precisely, the probability of *A* is twice as likely as the probability of both *BA* and *BB*. Giving the probabilities their proper weight, rather than 2 : 1, we should have 3 : 1 as the appropriate division of the championship monies. This would speak well for solution 3. That is to say that the case *A* in solution 1 is covered in solution 3 by the cases *AB* and *AA*. In reality, the two cases *AB* and *AA* do not really exist, since in each of these cases *A* would have been the winner with five games won, thus obviating the need for a second game. However, the reason we included these fictitious games is to show the equality of the probabilities.

Solution 2 is based on a false analogy. The equal likelihood of each player to win a game should not lead to allowing the uncompleted series to determine the partitioning of the money.

THE MISTAKE OF NOT THINKING BEFORE COUNTING

Very often a problem situation seems so simple that we plunge right in without first thinking about a strategy to use. This impetuous beginning for the solution often leads to a less elegant solution, and more often than not a mistake, as opposed to a method that results from a bit of forethought.

Here are two examples of simple problems, where we can avoid the normal pitfalls or mistakes and that can be made even simpler by thinking before working on them.

Problem I: Find all pairs of prime numbers whose sum equals 999.

One approach is to begin by taking a list of prime numbers and trying various pairs to see if they obtain 999 for a sum. This is obviously very tedious as well as time consuming, and you would never be quite certain that you had considered all the prime-number pairs.

Let's use some logical reasoning to solve this problem. In order to obtain an odd sum (the number 999) for two numbers—prime or otherwise—exactly one of the numbers must be even. Since there is only one even prime, namely 2, there can be only one pair of primes whose sum is 999, and that pair is 2 and 997. This simple reasoning will surely avoid some common mistakes.

A second problem where preplanning or some orderly thinking makes sense is as follows:

Problem II: A palindrome is a number that reads the same forward and backward, such as 747 or 1,991. How many palindromes are there between 1 and 1,000, inclusive?

The traditional approach to this problem would be to attempt to write out all the numbers between 1 and 1,000, and then to see which ones are palindromes. However, this is a cumbersome and time-consuming task at best, and one could easily omit some of them, which would lead to a mistaken answer.

Let's see if we can look for a pattern to solve the problem in a more direct fashion. Consider the following listing of numbers.

Range	Number of Palindromes	Total Number of Palindromes
1–9	9	9
10–99	9	18
100–199	10	28
200–299	10	38
300–399	10	48

There is a pattern. There are exactly ten palindromes in each group of one hundred numbers (after 99). Thus there will be nine sets of ten, or ninety, plus the eighteen from numbers 1 to 99 (that is, $9 + 9 + 9 \cdot 10 = 108$), for a total of 108 palindromes between 1 and 1,000.

Another solution to this problem would involve organizing the data in a favorable way. Consider all the single-digit numbers (self-palindromes). There are nine such. There are also nine two-digit palindromes. The three-digit palindromes have nine possible "outside digits" and ten possible "middle digits," so there are ninety of these. In total, there are 108 palindromes between 1 and 1,000, inclusive. The motto is: Think first, and then begin a solution. That way mistakes can be avoided.

A GAMBLER'S MISTAKE

Consider the following situation, which can be faced by a gambler and is deceptively misleading. (You may want to even simulate it with a friend to see if your intuition bears out.)

You are offered a chance to play a game. The rules are simple. There are one hundred cards, face down. Fifty-five of the cards say "*win*" and forty-five of the cards say "*lose*." You begin with a bankroll of $10,000. You must bet one-half of your money on each card turned over, and you either win or lose that amount based on what the card says. At the end of the game, all cards have been turned over. How much money do you have at the end of the game? Your guess may well be mistaken!

The same principle as above applies here. It is obvious that you will win ten times more than you will lose, so it appears that you will end with more than $10,000. What is obvious is often wrong, and this is a good example. Let's say that you win on the first card; you now have $15,000. Then you lose on the second card; you now have $7,500. If you had first lost and then won, you would still have $7,500. So every time you win one and lose one, you lose one-fourth of your money. So you end up with:

$$10{,}000 \cdot \left(\frac{3}{4}\right)^{45} \cdot \left(\frac{3}{2}\right)^{10}.$$

This is \$1.38 when rounded off. Surprised?

We could also look at the situation as follows: Assume you have D dollars at some stage. A win will increase D to $\frac{3}{2}D$, and a loss will decrease it to $\frac{1}{2}D$. After one hundred games, we will have fifty-five increases and forty-five decreases; therefore, we will have

$$10{,}000 \cdot \left(\frac{3}{2}\right)^{55} \cdot \left(\frac{1}{2}\right)^{45} \approx 1.37616\ldots \approx 1.38 \text{ dollars.}$$

THE CONTROVERSIAL AND FAMOUS MISTAKE KNOWN AS THE MONTY HALL PROBLEM

Let's Make a Deal was a long-running television game show that featured a problematic situation, one that was very prone to getting a mistaken answer from even some of the sharpest minds. The game goes this way: A randomly selected audience member would come on stage and be presented with three doors. She would be asked to select one, hopefully the one behind which there was a car, not one of the other two doors—each of which had a goat behind it. If she selected the door with the car, it was hers to keep. There was only one wrinkle in this: after the contestant made her selection, the host, Monty Hall, would expose one of the two goats behind a not-selected door (leaving two doors still unopened) and ask the participant if she wanted to stay with her original selection (not yet revealed) or switch to the other unopened door. At this point, to heighten the suspense, the rest of the audience would shout out "stay" or "switch" with seemingly equal frequency. The question is what to do? Does it make a difference? If so, which is the better strategy to use here (i.e., which gives her the greater probability of winning)? Most people give a mistaken response. They say that it doesn't make any difference, since they believe there is a 50 percent chance of guessing the correct door.

Let us look at this now step-by-step. The right answer will gradually become clear. There are *two goats* and *one car* behind these doors. The contestant must try to get the car. Suppose she selected door number 3.

Monty opens one of the doors that she *did not* select and exposes a goat, as shown in figure 5.23.

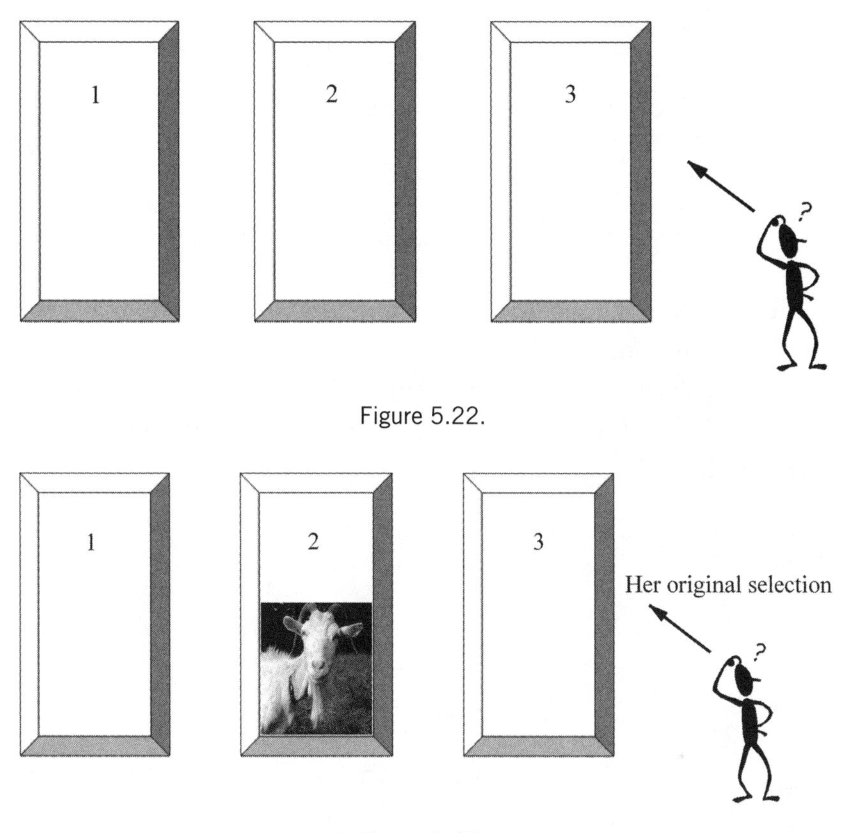

Figure 5.22.

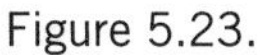

Figure 5.23.

Monty then asks: "Do you still want your first-choice door (No. 3), or do you want to switch to the other closed door?"

To help make the right decision, we shall consider an *extreme case*: Suppose there were one thousand doors instead of just three doors, and there are 999 goats and one car behind these doors.

Figure 5.24.

And suppose she chooses door number 1,000. How likely is it that she chose the right door?

"Very unlikely," since the probability of getting the right door is $\frac{1}{1{,}000}$.

How likely is it that the car is behind one of the other doors?

"Very likely": $\frac{999}{1{,}000}$.

Figure 5.25.

These doors (figure 5.25) are all "very likely" doors!

Monty Hall then opens all the doors (numbers 2 through 999) except one (say, door number 1), and he shows that each one had a goat, as shown in figure 5.26. A "very likely" door is left: door number 1.

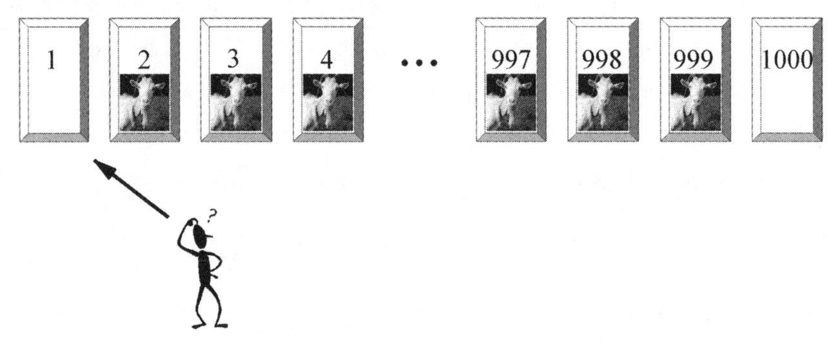

Figure 5.26.

We are now ready to answer the question. Which is a better choice?

Door number 1,000 ("very unlikely" door) or
Door number 1 ("very likely" door)?

The answer is now obvious. We ought to select the "very likely" door, which means "switching" is the better strategy for the audience participant to follow.

In the extreme case, it is much easier to see the best strategy than had we tried to analyze the situation with the three doors. The principle is the same in either situation.

This problem has caused many an argument in academic circles and was also a topic of discussion in the *New York Times* and in other popular publications as well. John Tierney wrote in the *New York Times* (Sunday, July 21, 1991):

> Perhaps it was only an illusion, but for a moment here it seemed that an end might be in sight to the debate raging among mathematicians, readers of *Parade* magazine, and fans of the television game show *Let's Make a Deal*. They began arguing last September after Marilyn vos Savant published a puzzle in *Parade*. As readers of her "Ask Marilyn" column are reminded each week, Ms. vos Savant is listed in the Guinness Book of World Records Hall of Fame for "Highest I.Q.," but that credential did not impress the public when she answered this question from a reader.

She gave the right answer, but still many mathematicians argued.

THE MISTAKE OF NOT CORRECTLY ANTICIPATING HEADS AND TAILS FROM A COIN TOSS

This lovely little situation will show you how some clever reasoning, along with algebraic knowledge of the most elementary kind, will help you solve a seemingly "impossibly difficult" problem—without mistake!

Consider the following problem:

> You are seated at a table in a dark room. On the table, there are twelve pennies, five of which are heads up and seven of which are tails up. (You know where the coins are, so you can move or flip any coin, but because it is dark, you will not know if the coin you are touching was originally heads up or tails up.) You are to separate the coins into two piles (and you are allowed to flip some of them if you want) so that when the lights are turned on, there will be an equal number of heads in each pile.

Your first reaction is, "You must be kidding! How can anyone do this task without seeing which coins are heads or tails up?" This just appears to be a mistake waiting to happen. However, this is where a most clever (yet incredibly simple) use of algebra will be the key to the solution.

Let's "cut to the chase." Here is what you do. (You might actually want to try it with twelve coins.) Separate the coins into two piles, of five and seven coins, respectively. Then flip over the coins in the smaller pile. Now both piles will have the same number of heads! That's all! You will think this is magic. How did this happen? Well, this is where algebra helps us understand what was actually done.

First, set twelve coins on a table such that there are five heads up and seven tails down. Then randomly select five coins for one pile and seven for another pile. When you separate the coins in the dark room, h heads will end up in the seven-coin pile. Then the other pile, the five-coin pile, will have $5 - h$ heads and $5 - (5 - h)$ tails. When you flip all the coins in the smaller pile, the $5 - h$ heads become tails and the $5 - (5 - h)$ tails become heads. Now each pile contains h heads!

THE MISTAKEN TEST RESULTS

We begin by assuming that in a population of one thousand, one person will have a specific disease. This can be determined by a test, where it is guaranteed that those with this disease will test positive. However, the test is not perfect, since some people will test positive even though they do not have the disease. In a population of one thousand, there will be ten who will test positive even though they do not have the disease. We need to determine what part of those tested will actually have the disease. Be aware that one of the solutions is mistaken.

Solution 1:

We know that of the population of one thousand there will be ten with a positive test result even though they do not have the disease. We also know that in the population of one thousand there will be one person having the disease. Therefore, eleven people will have tested positive, where only one will actually have the disease. Thus, the part that actually has the disease is $\frac{1}{11}$.

Solution 2:

For the solution we will consider a population of 100,000. In this population, every one-thousandth person is sick with the disease. That would indicate that there would be 100 people in this population with the disease. Therefore, the number of people who do not have the disease is 100,000 − 100 = 99,900. We know that among 1,000, there will be ten who will have tested positive but do not have the disease. That is, $\frac{10}{1{,}000} = \frac{1}{100}$ or 1 percent of the healthy people are incorrectly taken for diseased. One percent of 99,900 is 999 (since 99,900 ÷ 100 = 999). In total, then, there are 100 + 999 = 1,099 people with a positive result on the test. However, only 100 of these have the disease. Therefore, the truly diseased portion of the population is $\frac{100}{1{,}099}$.

Once again, our two solutions have two different results. Which one is the mistaken solution?

Yes, solution 2 is that correct one. In total, if one thousand people are tested, then there is one diseased person and the other 999 are not diseased. That is, of the healthy people, $\frac{10}{1{,}000} = \frac{1}{100}$ will erroneously receive a positive result. The test shows that of the 999 disease-free people, $\frac{999}{100} = 9.99$ disease-free people will show a positive test result. Instead of 9.99, we considered ten. That was the mistake. The $\frac{1}{100}$ relates only to the 999, and not on the full one thousand people.

We can also look at it from the point of view that of the one thousand people, exactly one is diseased.

The test actually shows that $1 + \frac{999}{100}$ people have a positive result. Therefore, the sought-after fraction is:

$$\frac{1}{1+\frac{999}{100}} = \frac{1}{\frac{1,099}{100}} = \frac{100}{1,099}.$$

With that, we would correct solution 1 and in effect have the result of solution 2. We merely used a population of 100,000 people so as to avoid decimals.

A COMMON MISTAKE OF ARRANGEMENTS: COLORING A STRIPED FLAG

The question we are asked here is, how many different striped flags can be created with four different colors at our disposal and with the flag having six horizontal stripes, with no two adjacent stripes of the same color?

Solution 1:

We will call the six stripes *A*, *B*, *C*, *D*, *E*, and *F*. Stripes *A*, *C*, and *E* are nonadjacent stripes have no restrictions, so they can be of any of the four colors. That means that there are four ways to fill each of these color slots.

A	4	
B		2
C	4	
D		2
E	4	
F		3

Since the stripe *B* cannot be the same color as *A* or *C*, there are only two possible colors for this stripe. The same is true for stripe *D*. However, stripe *F* can take on any color aside from that of stripe *E*, and so there are three possible colors that this stripe can take on. We show this in the chart above. Therefore, the number of possible stripe arrangements is $4 \cdot 2 \cdot 4 \cdot 2 \cdot 4 \cdot 3 = 768$, and so we can create 768 differently striped flags. This seems to be a reasonable solution. However, we have another solution to consider.

Solution 2:

This solution takes a different approach. Stripe A can take on a color in any one of four ways. Stripe B can then be colored in any one of three ways, so as not to match stripe A. The same argument then goes for each of the succeeding stripes. Using the multiplication principle, we get the following: $4 \cdot 3 \cdot 3 \cdot 3 \cdot 3 \cdot 3 = 972$. Thus, we can create 972 differently colored flags, as shown below.

A	4	
B		3
C		3
D		3
E		3
F		3

Once again, we are faced with a situation where our two apparently reasonable solutions give us two different answers. One of them must be a mistake.

As you might have expected, solution 1 is wrong. Although we were correct to say that the stripes A, C, and E can take on any of the four colors, the error lies in the fact that the next statement is simply inaccurate. The stripe B can have more possibilities than the two attributed to it, depending on the colors of stripes A and C. One must be careful not to be led into a reasonable-sounding mistaken solution.

WINNING A CONTEST BY AVOIDING A MISTAKEN STRATEGY

A team consists of three players. These three players stand in a circular arrangement and are blindfolded. While blindfolded, they will be given a hat to wear. The game coordinator, using a fair coin, places either a gray or a white hat on each of the players in the following way: If the coin shows heads, a gray hat is placed on the player's head. If the coin shows tails, a white hat is placed on the player's head. When the blindfolds are removed, each player can see the others' hats but not his own. A player can try to guess the color of his own hat, or he can say "I pass." The team wins if at least one player can guess the color of his own hat, and no player incorrectly guessed the color of his hat. Otherwise the team has lost. The players cannot speak with each other. Yet they can determine a common strategy to win the game. Which of the following strategies (solutions) would maximize the team's chances for winning?

Solution 1:

In order to win, the team needs to have the least one correct guess. The more responses they have, the greater their chances of winning. Therefore, the strategy will be for all three people to place a guess.

Solution 2:

In order not to lose, no player can select a false color. The more responses one gets, the greater the chance that someone will guess the wrong color hat on his head. In the event that no one speaks, then there will be no correct guess. Therefore, the optimum strategy here would be for only one player to guess the color of his own hat.

Solution 3:

It is possible for one, two, or three players to guess their own hat color. The more people who guess, the greater the chances of winning with correct guesses; however, the chances of someone mistakenly responding also increases. On the other hand, when fewer people respond with guesses, while probability of a false guess decreases, the probability of a correct guess also decreases. The ideal solution lies in the middle. The optimal strategy, then, is for exactly two people to guess.

These three strategies left us with three different results. Which one is correct?

Let's consider the three players with all the various possibilities: three players will be designated by A, B, and C. There are eight possibilities to consider (see figure 5.27).

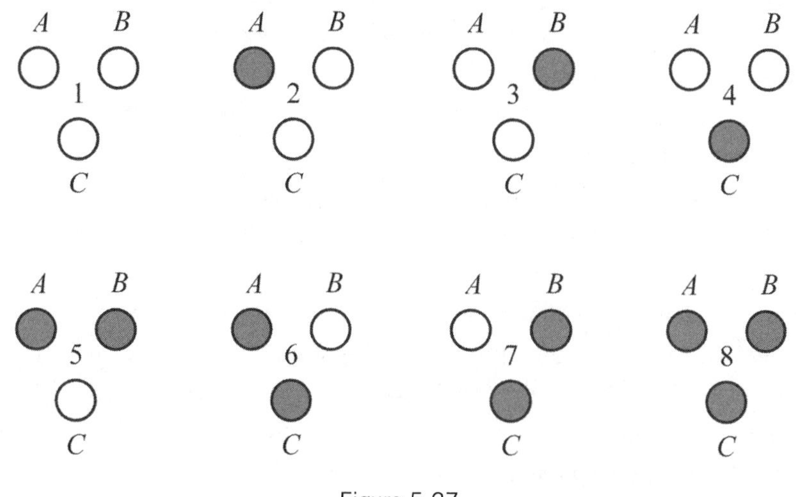

Figure 5.27.

Solution 1:

Let's assume that player A and player B both guessed a white hat, while player C guessed a gray hat. In that case, option 4 in figure 5.27 would be the only winning arrangement. Therefore, using this strategy, the probability of winning would be $\frac{1}{8}$ (= 12.5 percent).

Solution 2:

This time we shall assume that player B guessed that he has a gray hat and players A and C did not respond; in other words, they simply passed. The team will have won in cases 3, 5, 7, and 8, and they would have lost in all the other cases. Thus, the probability that the team would have won using this strategy is $\frac{4}{8}$ (= 50 percent).

Solution 3:

Now, we will assume that player B guesses he had a white hat, and that player C guessed he had a gray hat, while player A simply passed. In this case, the team would have won in options 4 and 6. Here the probability of the team winning is $\frac{2}{8}$ (= 25 percent).

How each of the players selects a hat color are independent events.

Solution 1 shows the lowest probability of winning, while solution 2 gives us the greatest chance of winning. It would then be advantageous to believe that solution 2 is the optimum solution. However, there are still better possibilities available.

Consider, then, the following fourth strategy that the team can use to win.

Right after the players remove their blindfolds, players A and B take note of the color of the hat on player C. If the hat on player C is white, then player A would immediately pass, or he would wait until after B passed. Player B does exactly the opposite. In this way, they can inform player C about the color of the hat on his head. This way, player C will be able to give the right answer about the color of the hat on his head. Here the probability of winning the game is 100 percent, which is clearly the best strategy for winning. Again we were enticed into accepting a reasonable-sounding strategy until we opened up to a more logical action.

AVOIDING THE MISTAKES IN DETERMINING THE RESULT OF TOSSING THREE DICE

Three dice are tossed at the same time. The first toss, A, produces a sum of eleven. The second toss, B, produces a sum of twelve. We are being asked here if the probability of event A is equal to the probability of event B. Let's consider the following solutions and see which one has a mistake and which one is correct.

Solution 1:

By listing the possible ways in which we can toss a sum of eleven or a sum of twelve, we can compare the two events clearly. We should note that the order of the dice is not important because of the laws of commutativity. We notice from the chart below that each of these sums can be obtained in six different ways. This would lead us to the conclusion that tossing a sum of eleven has the same likelihood as tossing a sum of twelve.

Sum of 11:	**Sum of 12:**
1 + 4 + 6	1 + 5 + 6
1 + 5 + 5	2 + 4 + 6
2 + 3 + 6	2 + 5 + 5
2 + 4 + 5	3 + 3 + 6
3 + 3 + 5	3 + 4 + 5
3 + 4 + 4	4 + 4 + 4

Solution 2:

Figures 5.28 and 5.29 show us the probability of tossing each one of the options that leads to the required sum. [We use the notation that P(*A*) indicates the probability of getting event *A*.]

P(*A*) =

$$\underset{1\,4\,6}{\tfrac{1}{6}\cdot\tfrac{1}{6}\cdot\tfrac{1}{6}}+\underset{1\,5\,5}{\tfrac{1}{6}\cdot\tfrac{1}{6}\cdot\tfrac{1}{6}}+\underset{1\,6\,4}{\tfrac{1}{6}\cdot\tfrac{1}{6}\cdot\tfrac{1}{6}}+\underset{2\,3\,6}{\tfrac{1}{6}\cdot\tfrac{1}{6}\cdot\tfrac{1}{6}}+\underset{2\,4\,5}{\tfrac{1}{6}\cdot\tfrac{1}{6}\cdot\tfrac{1}{6}}+\underset{2\,5\,4}{\tfrac{1}{6}\cdot\tfrac{1}{6}\cdot\tfrac{1}{6}}+\underset{2\,6\,3}{\tfrac{1}{6}\cdot\tfrac{1}{6}\cdot\tfrac{1}{6}}$$

$$+\underset{3\,2\,6}{\tfrac{1}{6}\cdot\tfrac{1}{6}\cdot\tfrac{1}{6}}+\underset{3\,3\,5}{\tfrac{1}{6}\cdot\tfrac{1}{6}\cdot\tfrac{1}{6}}+\underset{3\,4\,4}{\tfrac{1}{6}\cdot\tfrac{1}{6}\cdot\tfrac{1}{6}}+\underset{3\,5\,3}{\tfrac{1}{6}\cdot\tfrac{1}{6}\cdot\tfrac{1}{6}}+\underset{3\,6\,2}{\tfrac{1}{6}\cdot\tfrac{1}{6}\cdot\tfrac{1}{6}}+\underset{4\,1\,6}{\tfrac{1}{6}\cdot\tfrac{1}{6}\cdot\tfrac{1}{6}}+\underset{4\,2\,5}{\tfrac{1}{6}\cdot\tfrac{1}{6}\cdot\tfrac{1}{6}}$$

$$+\underset{4\,3\,4}{\tfrac{1}{6}\cdot\tfrac{1}{6}\cdot\tfrac{1}{6}}+\underset{4\,4\,3}{\tfrac{1}{6}\cdot\tfrac{1}{6}\cdot\tfrac{1}{6}}+\underset{4\,5\,2}{\tfrac{1}{6}\cdot\tfrac{1}{6}\cdot\tfrac{1}{6}}+\underset{4\,6\,1}{\tfrac{1}{6}\cdot\tfrac{1}{6}\cdot\tfrac{1}{6}}+\underset{5\,1\,5}{\tfrac{1}{6}\cdot\tfrac{1}{6}\cdot\tfrac{1}{6}}+\underset{5\,2\,4}{\tfrac{1}{6}\cdot\tfrac{1}{6}\cdot\tfrac{1}{6}}+\underset{5\,3\,3}{\tfrac{1}{6}\cdot\tfrac{1}{6}\cdot\tfrac{1}{6}}$$

$$+\underset{5\,4\,2}{\tfrac{1}{6}\cdot\tfrac{1}{6}\cdot\tfrac{1}{6}}+\underset{5\,5\,1}{\tfrac{1}{6}\cdot\tfrac{1}{6}\cdot\tfrac{1}{6}}+\underset{6\,1\,4}{\tfrac{1}{6}\cdot\tfrac{1}{6}\cdot\tfrac{1}{6}}+\underset{6\,2\,3}{\tfrac{1}{6}\cdot\tfrac{1}{6}\cdot\tfrac{1}{6}}+\underset{6\,3\,2}{\tfrac{1}{6}\cdot\tfrac{1}{6}\cdot\tfrac{1}{6}}+\underset{6\,4\,1}{\tfrac{1}{6}\cdot\tfrac{1}{6}\cdot\tfrac{1}{6}}=\frac{1}{8}=0.125$$

Figure 5.28.

$P(B) =$

$$\underset{1\;5\;6}{\underset{\uparrow\uparrow\uparrow}{\frac{1}{6}\cdot\frac{1}{6}\cdot\frac{1}{6}}}+\underset{1\;6\;5}{\underset{\uparrow\uparrow\uparrow}{\frac{1}{6}\cdot\frac{1}{6}\cdot\frac{1}{6}}}+\underset{2\;4\;6}{\underset{\uparrow\uparrow\uparrow}{\frac{1}{6}\cdot\frac{1}{6}\cdot\frac{1}{6}}}+\underset{2\;5\;5}{\underset{\uparrow\uparrow\uparrow}{\frac{1}{6}\cdot\frac{1}{6}\cdot\frac{1}{6}}}+\underset{2\;6\;4}{\underset{\uparrow\uparrow\uparrow}{\frac{1}{6}\cdot\frac{1}{6}\cdot\frac{1}{6}}}+\underset{3\;3\;6}{\underset{\uparrow\uparrow\uparrow}{\frac{1}{6}\cdot\frac{1}{6}\cdot\frac{1}{6}}}+\underset{3\;4\;5}{\underset{\uparrow\uparrow\uparrow}{\frac{1}{6}\cdot\frac{1}{6}\cdot\frac{1}{6}}}$$

$$+\underset{3\;5\;4}{\underset{\uparrow\uparrow\uparrow}{\frac{1}{6}\cdot\frac{1}{6}\cdot\frac{1}{6}}}+\underset{3\;6\;3}{\underset{\uparrow\uparrow\uparrow}{\frac{1}{6}\cdot\frac{1}{6}\cdot\frac{1}{6}}}+\underset{4\;2\;6}{\underset{\uparrow\uparrow\uparrow}{\frac{1}{6}\cdot\frac{1}{6}\cdot\frac{1}{6}}}+\underset{4\;3\;5}{\underset{\uparrow\uparrow\uparrow}{\frac{1}{6}\cdot\frac{1}{6}\cdot\frac{1}{6}}}+\underset{4\;4\;4}{\underset{\uparrow\uparrow\uparrow}{\frac{1}{6}\cdot\frac{1}{6}\cdot\frac{1}{6}}}+\underset{4\;5\;3}{\underset{\uparrow\uparrow\uparrow}{\frac{1}{6}\cdot\frac{1}{6}\cdot\frac{1}{6}}}+\underset{4\;6\;2}{\underset{\uparrow\uparrow\uparrow}{\frac{1}{6}\cdot\frac{1}{6}\cdot\frac{1}{6}}}$$

$$+\underset{5\;1\;6}{\underset{\uparrow\uparrow\uparrow}{\frac{1}{6}\cdot\frac{1}{6}\cdot\frac{1}{6}}}+\underset{5\;2\;5}{\underset{\uparrow\uparrow\uparrow}{\frac{1}{6}\cdot\frac{1}{6}\cdot\frac{1}{6}}}+\underset{5\;3\;4}{\underset{\uparrow\uparrow\uparrow}{\frac{1}{6}\cdot\frac{1}{6}\cdot\frac{1}{6}}}+\underset{5\;4\;3}{\underset{\uparrow\uparrow\uparrow}{\frac{1}{6}\cdot\frac{1}{6}\cdot\frac{1}{6}}}+\underset{5\;5\;2}{\underset{\uparrow\uparrow\uparrow}{\frac{1}{6}\cdot\frac{1}{6}\cdot\frac{1}{6}}}+\underset{5\;6\;1}{\underset{\uparrow\uparrow\uparrow}{\frac{1}{6}\cdot\frac{1}{6}\cdot\frac{1}{6}}}+\underset{6\;1\;5}{\underset{\uparrow\uparrow\uparrow}{\frac{1}{6}\cdot\frac{1}{6}\cdot\frac{1}{6}}}$$

$$+\underset{6\;2\;4}{\underset{\uparrow\uparrow\uparrow}{\frac{1}{6}\cdot\frac{1}{6}\cdot\frac{1}{6}}}+\underset{6\;3\;3}{\underset{\uparrow\uparrow\uparrow}{\frac{1}{6}\cdot\frac{1}{6}\cdot\frac{1}{6}}}+\underset{6\;4\;2}{\underset{\uparrow\uparrow\uparrow}{\frac{1}{6}\cdot\frac{1}{6}\cdot\frac{1}{6}}}+\underset{6\;5\;1}{\underset{\uparrow\uparrow\uparrow}{\frac{1}{6}\cdot\frac{1}{6}\cdot\frac{1}{6}}}=\frac{25}{216}\approx 0.116$$

Figure 5.29.

This can be summed up in the following table.

Sum of 11:	Sum of 12:
1 + 4 + 6 is obtained 6 times	1 + 5 + 6 is obtained 6 times
1 + 5 + 5 is obtained 3 times	2 + 4 + 6 is obtained 6 times
2 + 3 + 6 is obtained 6 times	2 + 5 + 5 is obtained 3 times
2 + 4 + 5 is obtained 6 times	3 + 3 + 6 is obtained 3 times
3 + 3 + 5 is obtained 3 times	3 + 4 + 5 is obtained 6 times
3 + 4 + 4 is obtained 3 times	4 + 4 + 4 is obtained 1 times

There are twenty-seven cases that sum up to eleven and twenty-five cases that sum up to twelve.

A comparison of the two probabilities shows that the two events, tossing a sum of eleven or tossing a sum of twelve, are almost equally likely.

Solution 3:

If we consider the possible sums we can get from the three dice, we can notice where the sums of eleven and twelve would be along the list:

3; 4; 5; 6; 7; 8; 9; 10 || **11**; **12**; 13; 14; 15; 16; 17; 18.

At the extreme ends we find that the sum of three can only be achieved in one way, as is the case for the sum of eighteen. That is, $3 = 1 + 1 + 1$, and $18 = 6 + 6 + 6$.

The possibilities to get sums four and seventeen are as follows: $4 = 1 + 1 + 2 = 1 + 2 + 1 = 2 + 1 + 1$, and $17 = 6 + 6 + 5 = 6 + 5 + 6 = 5 + 6 + 6$.

In general, the number of ways each of the sums can be achieved are symmetric on both sides of the middle. This then translates to the notion that the probability of getting the sum of ten is the same as the probability of getting the sum of eleven. Similarly, the following probabilities of obtainable sums are also equal: $P(9) = P(12)$, and $P(4) = P(17)$.

This would indicate that the probability of getting a sum of eleven is not equal to that of getting a sum of twelve.

We are once again faced with the dilemma of having three solutions with different results, yet each seems somewhat reasonable. There must be a mistake somewhere—but where?

Solution 1 is in error. Although the six possibilities for getting the sums of eleven and twelve are correct, the conclusion is simply wrong. The likelihood of each of the sums is important, as we show in figure 5.30.

P("3 + 3 + 6") =	$\underset{3}{\frac{1}{6}} \cdot \underset{3}{\frac{1}{6}} \cdot \underset{6}{\frac{1}{6}} + \underset{3}{\frac{1}{6}} \cdot \underset{6}{\frac{1}{6}} \cdot \underset{3}{\frac{1}{6}} + \underset{6}{\frac{1}{6}} \cdot \underset{3}{\frac{1}{6}} \cdot \underset{3}{\frac{1}{6}} = \frac{3}{216}$
P("4 + 4 + 4") =	$\underset{4}{\frac{1}{6}} \cdot \underset{4}{\frac{1}{6}} \cdot \underset{4}{\frac{1}{6}} = \frac{1}{216}$

Figure 5.30.

Further sum-probabilities for the three-dice sums of eleven and twelve are shown below, and we notice that in some cases the probabilities are equal—but not in all.

$$P(1+4+6) = P(1+5+6) = \frac{6}{216},$$

$$P(2+4+5) = P(3+4+5) = \frac{6}{216},$$

$$P(1+5+5) = P(2+5+5) = \frac{3}{216},$$

$$P(3+3+5) = P(3+3+6) = \frac{3}{216},$$

$$P(2+3+6) = P(2+4+6) = \frac{6}{216}, \text{ but for}$$

$P(3+4+4)$ and $P(4+4+4)$, the probabilities are not equal since $\frac{3}{216} \neq \frac{1}{216}$.

Five of the pairs of probabilities are equal to each other, as indicated in the first five lines above. However, $P(3+4+4) > P(4+4+4)$, which implies that the $P(11) > P(12)$.

There is also a problem with solution 2, in that the combination (1, 6, 5) was used instead of 1, 6, 4 to get the sum of eleven. Additionally, the following were also excluded (2, 6, 3), (3, 5, 3), and (3, 6, 2). Therefore, the correct probability for A is $P(A) = \frac{27}{216} = 0.125$.

Thus, since $0.1157 \ldots \neq 0.125$, the two probabilities are not equal. Solution 3 is correct, since eleven is closer to the middle point of the sequence than is twelve, and it follows that $P(11) > P(12)$.

THE MISTAKE TO AVOID TO DECREASE ONE'S CHANCES OF LOSING!

The manufacturer of a game machine that contains countless non-transparent balls fills every fifth ball with a $5.00 bill; all of the other balls are filled with a blank piece of paper the same size as a $5.00 bill. If one selects three balls randomly all at one time, what is the probability, P(E), that none of the three balls will contain a $5.00 bill?

Solution 1:

We will begin by letting F represent the $5.00 bill and X represent the blank piece of paper. Therefore, $P(F) = \frac{1}{5}$ and $P(X) = \frac{4}{5}$.

The probability of not selecting the $5.00 bill on three attempts can be represented by:

$$P(E) = \underset{X}{\frac{4}{5}} \cdot \underset{X}{\frac{4}{5}} \cdot \underset{X}{\frac{4}{5}} = 0.512.$$

Solution 2:

We will let ● represent a ball containing a $5.00 bill, and o will represent the ball containing the blank piece of paper. In other words, we are going to work with the combination o o o o ● taken three at a time. There are $\binom{5}{3} = \frac{5 \cdot 4 \cdot 3}{1 \cdot 2 \cdot 3} = 10$ ways that the three balls can be selected. If we consider only the four o-balls, then there are $\binom{4}{3} = \frac{4 \cdot 3 \cdot 2}{1 \cdot 2 \cdot 3} = 4$ ways they can be selected, which is the group of four balls that we originally chose to be as our target group.

From the definition of probability, the probability of obtaining three balls without a $5.00 bill is $P(E) = \frac{4}{10} = 0.4$.

Once again, we find ourselves with two solutions and two different answers. Therefore, one of the solutions must be in error.

Solution 1 is correct. Since the balls are filled at random, yet every fifth ball is filled with a $5.00 bill, it is clear that the probability of getting a ball with a \$5.00 bill is $\frac{1}{5}$. Of course, then a probability of getting a ball with a blank piece of paper is $\frac{4}{5}$.

Therefore, solution 2 must be mistaken. The reason is that solution 2 is based on the notion that exactly five balls, one of which contained a \$5.00 bill, were selected and from these, three balls were picked.

As there were a large number of balls produced, when the ball is selected, the probability of success and failure is still $\frac{1}{5}$ and $\frac{4}{5}$, respectively.

However, in the case where exactly five balls are at hand, we can calculate the probability of selecting three balls without a \$5.00 bill as:

$$P(E) = \frac{4}{5} \cdot \frac{3}{4} \cdot \frac{2}{3} = 0.4.$$

$$X \quad X \quad X$$

AVOIDING A MISTAKE IN GETTING THE CORRECT TELEPHONE NUMBER

We find that David has forgotten the last digit of a telephone number. As he tries to determine the correct telephone number, what is the probability, $P(E)$, that he will get the correct last digit on not more than two tries?

Solution 1:

The numbers 0, 1, 2, 3, 4, 5, 6, 7, 8, and 9 are the ten digits we can dial. With these digits we can perform $10 \cdot 10 = 100$ pairs, including repetition of digits ("stupid tries"). We assume that all these pairs are of equal probability. Let us consider the case that the last digit of the phone number is 1. Then we have nineteen pairs where this digit occurs:

These are the ten cases where the first try is successful: 10, 11, 12, 13, 14, 15, 16, 17, 18, 19.

These are the nine cases where first try does not match, but the second try is successful: 01, 21, 31, 41, 51, 61, 71, 81, 91.

Dialing the digit 1 with at most two tries will therefore have a probability of $\frac{19}{100} = 0.19$.

Solution 2:

We know that 0, 1, 2, 3, 4, 5, 6, 7, 8, and 9 are the ten digits we can dial. With these digits we can perform $90 = 10 \cdot 9$ pairs without repetition of digits ("intelligent tries"). We assume that all these pairs are of equal probability. Let us consider the case that the last digit of the phone number is 1. Then we have eighteen pairs where this digit occurs:

These are the nine cases where the first try is successful: 10, 12, 13, 14, 15, 16, 17, 18, 19.

These are the nine cases where first try does not match, but the second try is successful: 01, 21, 31, 41, 51, 61, 71, 81, 91.

Selecting digit 1 with at most two tries will therefore have a probability of $\frac{18}{90} = 0.2$.

Solution 2 is correct, but only by 1 percent. Although it's a small difference, it's nevertheless a mistake in solution 1.

An additional question with which we could challenge ourselves is, what is the probability that David does not need more than three tries? The answer should now be clear as the following:

$$P(E) = \frac{1}{10} + \frac{9}{10} \cdot \frac{1}{9} + \frac{9}{10} \cdot \frac{8}{9} \cdot \frac{1}{8} = 0.3.$$

THE COUNTERINTUITIVE PREDICTION OF THE SEX OF A CHILD

The parents of a son are about to have another child. What is the probability $P(E)$ that the second child will be a girl? (We will assume that it is equally likely to get a boy or a girl.)

Solution 1:

Since it is equally likely to get a boy or a girl, the probability would then be $\frac{1}{2} = 0.5$.

Solution 2:

This time we will work with the probability of not getting a girl. We will let B represent the birth of a boy and G represent giving birth to a girl.

Hence, the probability of not getting a girl is $P(\overline{G}) = \underset{B}{\frac{1}{2}} \cdot \underset{B}{\frac{1}{2}} = \frac{1}{4} = 0.25.$

Therefore, $P(G) = 1 - P(\overline{G}) = \frac{3}{4} = 0.75$. Therefore, the probability of getting a girl is then $\frac{3}{4} = 0.75$.

Solution 3:

Here we will let G represent the girl and B represent the boy. Therefore, the possibilities are as follows: BB, BG, and GB—where in each case a boy exists. Since there are two cases out of three where a girl appears, the probability of their getting a girl is $\frac{2}{3} \approx 0.66667$.

Solution 4:

There are three possibilities for these parents: two boys, two girls, and a boy and a girl. Since we already know that the parents have a boy, we can eliminate the one case of the two girls. From the remaining two cases, the only one that we would see as successful is the one of the boy and girl. Therefore, the probability is $\frac{1}{2} = 0.5$.

Solution 5:

This time we will take it case by case. We will simply calculate the probability of getting a boy and a girl, or a girl and a boy:

$$P(G) = \underset{B}{\frac{1}{2}} \cdot \underset{G}{\frac{1}{2}} + \underset{G}{\frac{1}{2}} \cdot \underset{B}{\frac{1}{2}} = \frac{1}{2} = 0.5.$$

As expected, we have various solutions and we must determine which one is correct, and find where our mistakes are in the other solutions.

Solutions 2 and 3 use false reasoning and yield false numbers. Solutions 1 and 4 have sound arguments and the correct probability. Solution 5 has false reasoning but the correct probability.

Solution 5 is an example of where only due to the simplicity of our problem, we sometimes get the correct number for the required probability by arguing mistakenly. This solution essentially solves correctly a different question. That question would be, what is the probability of a pair of siblings being of different sex? The difference between this question and that of the given one is that it does not assume that one of the children is already a boy—as was the case in the given question.

CONCLUSION

We hope you have enjoyed this journey through a myriad of mistakes in mathematics. Some mistakes were quite comical, others led famous mathematicians to struggle further toward profound results, and yet others allowed you to both appreciate the power and beauty of mathematics and realize how the structure was sometimes based on premises or approaches that led from mistakes to surprising successes.

For example, we know that division by zero is not permissible since it leads to absurd results, and yet we found mistakes embedded in mathematical procedures where it was not recognized that a division by zero had occurred. We also saw some mistakes in geometry that on occasion might have appeared to be misdrawn diagrams but were actually attributed to either a lack of precision, or the absence of a definition (such as of the word *betweenness*).

However we look at mistakes in mathematics, we hope that your journey through this variety of mistakes and errors will leave you with a more genuine appreciation of this most important science. For understanding mistakes in mathematics can help you become not only a better mathematician, but also a better thinker. The famous German mathematician Carl Friedrich Gauss (1777–1855) referred to mathematics as "the queen of sciences," and we want to make sure that readers understand that mistakes rectified help assure this special title.

NOTES

INTRODUCTION

1. See Alfred S. Posamentier and Ingmar Lehmann, *Pi: A Biography of the World's Most Mysterious Number*, with an afterword by Nobel laureate Herbert Hauptman (Amherst, NY: Prometheus Books, 2004), p. 70ff.

CHAPTER 1: NOTEWORTHY MISTAKES BY FAMOUS MATHEMATICIANS

1. For more on this amazing ratio, see Alfred S. Posamentier and Ingmar Lehmann, *The Glorious Golden Ratio* (Amherst, NY: Prometheus Books, 2012).

2. See the complete collection of logarithmic and trigonometric tables by Adrian Black in *Arithmetica Logarithmica* and *Trigonometria artificialis* (Leipzig 1794), improved and increased version.

3. Lutz Führer, "Geniale Ideen und ein lehrreicher Fehler des berühmten Herrn Galilei," *Mathematica didactica* 28, no 1 (2005): S. 58–78.

4. Greek: *brachistos* = shortest, *chronos* = time.

5. Alexandre Koyre, *Leonardo, Galilei, Pascal—Die Anfänge der neuzeitlichen Naturwissenschaft* (Frankfurt am Main: Fischer, 1998), p. 178.

6. Jakob Steiner, "Einige geometrische Sätze," *Journal reine angewandte Mathematik* 1(1826): 38–52; "Einige geometrische Betrachtungen," *Journal reine angewandte Mathematik* 1(1826): 161–84; and "Fortsetzung der geometrischen Betrachtungen," *Journal reine angewandte Mathematik* 1(1826): 252–88.

7. H. Lob and H. W. Richmond, "On the Solutions of Malfatti's Problem for a Triangle," *Proceedings London Mathematical Society* 2, no. 30 (1930): 287–301.

8. Michael Goldberg, "On the Original Malfatti Problem," *Mathematics Magazine* 40 (1967): 241–47.

9. Malfatti misconstrued his own problem. (Richard K. Guy, "The Lighthouse Theorem, Morley & Malfatti—A Budget of Paradoxes," *American Mathematics Monthly* 114, no. 2 (2007): 97–141).

10. V. A. Zalgaller and G. A. Los, "The Solution of Malfatti's Problem," *Journal of Mathematical Sciences* 72, no. 4 (1994): 3163–77.

11. M. Gardner, "Mathematical Games: Six Sensational Discoveries That Somehow or Another Have Escaped Public Attention," *Scientific American* 232 (April 1975): 127–31; M. Gardner, "Mathematical Games: On Tessellating the Plane with Convex Polygons," *Scientific American* 232 (July 1975): 112–17.

12. Kenneth Appel and Wolfgang Haken, "The Solution of the Four-Color-Map Problem," *Scientific American* 237, no. 4 (1977): 108–21.

13. Preda Mihăilescu, "Primary Cyclotomic Units and a Proof of Catalan's Conjecture," *Journal Reine Angewandte Mathematik* 572 (2004): 167–95.

14. See Alfred S. Posamentier and Ingmar Lehmann, *The Fabulous Fibonacci Numbers* (Amherst, NY: Prometheus Books, 2007).

15. Tomás Oliveira e Silva, "Goldbach Conjecture Verification: Introduction," last updated November 22, 2012, http://www.ieeta.pt/~tos/goldbach.html (accessed April 15, 2013); Alfred S. Posamentier and Ingmar Lehmann, *Mathematical Amazements and Surprises: Fascinating Figures and Noteworthy Numbers*, with an afterword by Nobel laureate Herbert Hauptman (Amherst, NY: Prometheus Books, 2009), p. 226.

16. See Tomás Oliveira e Silva, "Computational Verification of the *3x+1* Conjecture," http://www.ieeta.pt/~tos/3x+1.html. Also see Tomás Oliveira e Silva, "Maximum Excursion and Stopping Time Record-Holders for the 3x+1 Problem: Computational Results," *Mathematics of Computation* 68, no. 225 (1999):371–84.

17. Posamentier and Lehmann, *Mathematical Amazements and Surprises*, pp. 111–26.

18. Martin Aigner and Günter M. Ziegler, *Proofs from THE BOOK* (Berlin: Springer, 1998), pp. 7–12.

19. D. B. Gillies, "Three New Mersenne Primes and a Statistical Theory," *Mathematics Computing* 18 (1964): 93–97.

20. P. Ochem and M. Rao, "Odd Perfect Numbers Are Greater Than 10^{1500}," *Mathematics of Computation* (2011), http://www.lirmm.fr/~ochem/opn/opn.pdf.

21. Paulo Ribenboim, *The Little Book of Bigger Primes* 2nd ed. (New York: Springer, 2004).

22. See Posamentier and Lehmann, *Mathematical Amazements and Surprises*, p. 10.

23. A. de Polignac, "Six Propositions Arithmologiques d'Eduites de Crible d'Eratosthène," *Nouvelles Annales de Mathématiques* 8 (1849): 423–29.

24. R. Crocker, "On the Sum of a Prime and Two Powers of Two," *Pacific Journal of Mathematics* 36 (1971): 103–107.

25. L. J. Lander and T. R. Parkin, "Counterexample to Euler's Conjecture on Sums of Like Powers," *Bulletin of the American Mathematical Society* 72 (1966): 1079; and

Leon J. Lander, Thomas R. Parkin, and John L. Selfridge, "A Survey of Equal Sums of Like Powers," *Mathematics of Computation* 21 (1967): 446–59.

26. Noam D. Elkies, "On $A^4+B^4+C^4=D^4$," *Mathematics of Computation* 51 (1988): 825–35.

27. Gaston Tarry, "Le Probléme de 36 Officiers," *Compte Rendu de l'Association Française pour l'Avancement de Science Naturel* 1 (Secrétariat de l'Association) (1900): 122–23.

28. H. F. Mac Neish, "Euler Squares," *Annals of Mathematics* 23(1922): 221–27.

29. R. C. Bose and S. S. Shrikhande, "On the Falsity of Euler's Conjecture about the Nonexistence of Two Orthogonal Latin Squares of Order 4t+2," *Proceedings of the National Academy of Science* 45 (1959): 734–37.

30. E. T. Parker, "Construction of Some Sets of Mutually Orthogonal Latin Squares," *Proceedings of the American Mathematical Society* 10 (1959): 946–49; and E. T. Parker, "Orthogonal Latin Squares," *Proceedings of the National Academy of Science* USA 45 (1959): 859–62.

31. R. C. Bose, S. S. Shrikhande, and E. T. Parker, "Further Results on the Construction of Mutually Orthogonal Latin Squares and the Falsity of Euler's Conjecture," *Canadian Journal of Mathematics* 12 (1960): 189–203.

32. V. I. Ivanov, "On Properties of the Coefficients of the Irreducible Equation for the Partition of the Circle," *Uspekhi Matematicheskikh Nauk* 9 (1941): 313–17.

33. Hans Ohanian, *Einstein's Mistakes: The Human Failings of Genius*, 1st ed. (New York: W. W. Norton, 2008).

CHAPTER 2: MISTAKES IN ARITHMETIC

1. See Mike Sutton, "Spinach, Iron, and Popeye: Ironic Lessons from Biochemistry and History on the Importance of Healthy Eating, Healthy Skepticism and Adequate Citation," *Internet Journal of Criminology* (2010): 1–34, http://www.internetjournalofcriminology.com/Sutton_Spinach_Iron_and_Popeye_March_2010.pdf.

2. C. Stanley Ogilvy and John T. Anderson, *Excursions in Number Theory* (New York: Oxford University Press, 1966), p. 86.

3. A. P. Darmoryad, *Mathematical Games and Pastimes* (New York: Macmillan, 1964), p. 35.

4. Raphael Robinson, "C. W. Trigg: E 69," *American Mathematical Monthly* 41, no. 5 (1934): 332.

CHAPTER 3: ALGEBRAIC MISTAKES

1. This limit is obtained by the Taylor series $\ln(1+x) = \frac{x}{1} - \frac{x^2}{2} + \frac{x^3}{3} - \frac{x^4}{4} \pm \ldots$.

2. The natural logarithm is the exponent of the base $e \approx 2.718$, whereas the common logarithm is the exponent of the base 10.

3. Compare with Gottfried Wilhelm Leibniz's mistake, mentioned in chapter 1.

4. "Gleanings Far and Near," *Mathematical Gazette* 33, no. 2 (1949): 112.

5. A. G. Konforowitsch, *Logischen Katastrophen auf der Spur* (Leipzig: Fachbuchverlag, 1997), p. 83.

6. Alfred S. Posamentier and Ingmar Lehmann, *The Fabulous Fibonacci Numbers*, with an afterword by Nobel laureate Herbert Hauptman (Amherst, NY: Prometheus Books, 2007), pp. 78–81.

CHAPTER 4: GEOMETRIC MISTAKES

1. These so-called Müller-Lyer illusions were developed in 1889 by the German psychiatrist Franz Müller-Lyer (1857–1916).

2. Swedish postage stamp: 25 Öre, Sverige, February 16, 1982.

3. More such examples can be found in Alfred S. Posamentier and Ingmar Lehmann, *The Fabulous Fibonacci Numbers*, with an afterword by Nobel laureate Herbert Hauptman (Amherst, NY: Prometheus Books, 2007), pp. 140–43.

4. The angle bisector of a triangle divides the side to which it is drawn proportional to the two adjacent sides. See Alfred S. Posamentier and Ingmar Lehmann, *The Secrets of Triangles* (Amherst, NY: Prometheus Books, 2012), p. 43.

5. Berthold Schuppar and Hans Humenberger, "Drachenvierecke mit einer besonderen Eigenschaft," *Math. Naturwiss. Unterricht* 60, no. 3 (2007): 140–45.

6. Circumference $= 2\pi \cdot r$; using higher mathematics, we can show that a regular cycloid has length $= 8r$, when r is the radius of the circle that has made one complete revolution.

7. Apparently the first publication of this "classic" problem appeared in the article "The Paradox Party. A Discussion of Some Queer Fallacies and Brain-Twisters" by Henry Ernest Dudeney. *Strand Magazine* 38, no. 228, ed. George Newnes (December 1909): 670–76.

8. For more cases and a discussion of similar problems, see Alfred S. Posamentier and Ingmar Lehmann, *Pi: A Biography of the World's Most Mysterious Number*, with an afterword by Nobel laureate Herbert Hauptman (Amherst, NY: Prometheus Books, 2004), pp. 222–43, 305–308.

9. David A. James, Ian Richards, David E. Kullman, and Lyman C. Peck, "News

and Letters," *Mathematics Magazine* 54, no. 3 (1981): 148–53; see also the March 31, 1981, issue of *Time*, (p. 51) or the April 6, 1981, issue of *Newsweek* (p. 84). See also our introduction.

10. Gustav Fölsch, "Haben Schüler einen sechsten Sinn?" *Praxis der Mathematik* 26, no. 7 (1984): 211–15.

BIBLIOGRAPHY

Ball, W.W. Rouse. *Mathematical Recreations and Essays*. New York: Macmillan, 1960.

Barbeau, Edward J. *Mathematical Fallacies, Flaws, and Flimflam*. Washington, DC: Mathematical Association of America, 2000.

Bunch, Bryan H. *Mathematical Fallacies and Paradoxes*. New York: Van Nostrand Reinhold, 1982.

Campbell, Stephen K. *Flaws and Fallacies in Statistical Thinking*. Englewood Cliffs, NJ: Prentice-Hall, 1974.

Darmoryad, A. P. *Mathematical Games and Pastimes*. New York: Macmillan, 1964.

Dubnov, Ya. S. *Mistakes in Geometric Proofs*. Boston: D. C. Heath, 1963.

Dudeney, H. E. *Amusements in Mathematics*. New York: Dover, 1970.

Furdek, Atilla, *Fehler-Beschwörer—Typische Fehler beim Lösen von Mathematikaufgaben*. Norderstedt: Books on Demand, 2002.

Gardner, Martin. *Fads and Fallacies: In the Name of Science*. New York: Dover, 1957.

———. *Perplexing Puzzles and Tantalizing Teasers*. New York: Dover, 1988.

Havil, Julian. *Impossible?—Surprising Solutions to Counterintuitive Conundrums*. Princeton, NJ: Princeton University, 2008.

———. *Nonplussed!—Mathematical Proof of Implausible Ideas*. Princeton, NJ: Princeton University, 2007.

James, Ioan. *Remarkable Mathematicians*. Cambridge: Cambridge University, 2002.

Jargocki, Christopher P. *Science Brain-Twisters, Paradoxes, and Fallacies*. New York: Charles Scribner, 1976.

Konforowitsch, Andrej G. *Logischen Katastrophen auf der Spur*. Leipzig: Fachbuchverlag, 1992.

Kracke, Helmut. *Mathe-musische Knobelisken*. Bonn: Dümmler, 1983.

Lietzmann, Walther. *Wo steckt der Fehler?* Leipzig: Teubner, 1952.

Madachy, Joseph. *Mathematics on Vacation*. New York: Charles Scribner, 1966.

Maxwell, E. A. *Fallacies in Mathematics*. London: Cambridge University Press, 1959.

Northrop, Eugene P. *Riddles in Mathematics*. Princeton, NJ: D. Van Nostrand, 1944.

O'Beirne, T. H. *Puzzles and Paradoxes*. New York: Oxford University, 1965.

Posamentier, Alfred S. *Advanced Euclidean Geometry*. Hoboken, NJ: Wiley, 2002.

———. *The Pythagorean Theorem: The Story of Its Power and Glory*. Afterword by Nobel laureate Herbert Hauptman. Amherst, NY: Prometheus Books, 2010.

Posamentier, Alfred S., and Ingmar Lehmann. *Mathematical Amazements and Surprises:*

Fascinating Figures and Noteworthy Numbers. Afterword by Nobel laureate Herbert Hauptman. Amherst, NY: Prometheus Books, 2009.

Schumer, Peter D. *Mathematical Journeys*. Hoboken, NJ: Wiley, 2004.

Scripta Mathematica 5 (1938) through 12 (1946).

White, William F. *A Scrapbook of Elementary Mathematics*. LaSalle, IL: Open Court, 1942.

Wurzel. *Zeitschrift für Mathematik* (Jena, Germany) 38 (2004) through 46 (2012).

INDEX